T0139142

INTERNET OF THINGS

INTERNET OF THINGS
Energy, Industry, and Healthcare

Edited by
Arun Kumar Rana, Ayodeji Olalekan Salau, Sharad Sharma, Shubham Tayal, and Swati Gupta

CRC Press
Taylor & Francis Group
Boca Raton London New York

CRC Press is an imprint of the
Taylor & Francis Group, an **informa** business

First edition published 2022
by CRC Press
2 Park Square, Milton Park, Abingdon, Oxon, OX14 4RN

and by CRC Press
6000 Broken Sound Parkway NW, Suite 300, Boca Raton, FL 33487-2742

© 2022 selection and editorial matter, Arun Kumar Rana, Ayodeji Olalekan Salau,
Sharad Sharma, Shubham Tayal, Swati Gupta; individual chapters, the contributors

CRC Press is an imprint of Informa UK Limited

Library of Congress Cataloging-in-Publication Data
Names: Rana, Arun Kumar, editor.
Title: Internet of things: energy, industry, and healthcare / edited by Arun Kumar Rana,
Ayodeji Olalekan Salau, Sharad Sharma, Shubham Tayal, Swati Gupta.
Other titles: Internet of things (Rana)
Description: First edition. | Abingdon, Oxon; Boca Raton, FL: CRC Press, 2022. | Includes
bibliographical references and index. | Summary: "This book provides future research
directions in energy, industry, and healthcare domain, and explores the different
applications of IoT and its associated technologies. The book discusses software
architecture, middleware, data processing and management as well as security, sensors and
algorithms" – Provided by publisher.
Identifiers: LCCN 2021019165 (print) | LCCN 2021019166 (ebook) | ISBN
9780367686529 (hb) | ISBN 9780367691097 (pbk) | ISBN 9781003140443 (ebk)
Subjects: LCSH: Internet of things.
Classification: LCC TK5105.8857.I566 2022 (print) | LCC TK5105.8857 (ebook) | DDC
004.67/8–dc23
LC record available at https://lccn.loc.gov/2021019165
LC ebook record available at https://lccn.loc.gov/2021019166

ISBN: 978-0-367-68652-9 (hbk)
ISBN: 978-0-367-69109-7 (pbk)
ISBN: 978-1-003-14044-3 (ebk)

DOI: 10.1201/9781003140443

Typeset in Times
by MPS Limited, Dehradun

Contents

Chapter 10 A Robust Context and Role-Based Dynamic Access Control
for Distributed Healthcare Information Systems............................131

*Abdulkadir Abdulkadir Adamu, Ayodeji Olalekan Salau, and
Li Zhiyong*

Chapter 11 Impact of ICT on Handicrafts Marketing in Delhi
NCR Region 153

Archita Nandi

Chapter 12   Intelligent Amalgamation of Blockchain Technology with
                  Industry 4.0 to Improve Security.....................................165

Sumit Kumar Rana and Sanjeev Kumar Rana

Chapter 13   Sensor Networks and Internet of Things in Agri-Food.................177

Moses Oluwafemi Onibonoje

Chapter 14   Design and Development of Hybrid Algorithms to Improve
                  Cyber Security and Provide Securing Data Using Image
                  Steganography with Internet of Things 195

Abhishek Mehta and Trupti Rathod

Chapter 18 Smart Technologies and Social Impact: An Indian Perspective
of Contactless Technologies for Pandemic 275

*Rashmi Bhardwaj, Varsha Duhoon, and
Mohammad Ayoub Khan*

Chapter 19 Agriculture-Internet of Things (A-IoT) Key Roles in
Addressing Some Challenges in Agriculture 301

Ibrahim Muhammad Abdul

Editor Biographies

Prof. Arun Kumar Rana completed his B.Tech degree from Kurukshetra University and M.Tech. Degree from Maharishi Markandeshwar (Deemed to be University), in Mullana, India. Mr. Rana is currently pursuing his Ph.D. from Maharishi Markandeshwar (Deemed to be University), in Mullana, India. His areas of interest include Image Processing, WSN, IoT, AI, and Machine Learning and Embedded systems. Mr. Rana is currently working as an Assistant Professor in PIET College Samalkha, with more than 13 years of experience.

Dr. Ayodeji Olalekan Salau received a B.Eng. in Electrical/Computer Engineering from the Federal University of Technology, in Minna, Nigeria in 2010. He received his M.Sc and Ph.D. from the Obafemi Awolowo University, in Ile-Ife, Nigeria in 2015 and 2018, respectively. His research interests include computer vision, image processing, signal processing, machine learning, power systems technology, and nuclear engineering. He is a registered Engineer with the Council for the Regulation of Engineering in Nigeria (COREN), a member of the International Association of Engineers (IAENG), and a recipient of the Quarterly Franklin Membership with ID number CR32878 given by the Editorial Board of London Journals Press in 2020.

Dr. Sharad Sharma has vast experience of more than 20 years of teaching and administrative work. He has received his B. Tech degree in Electronics Engineering from Nagpur University, in Nagpur, India, M.Tech in Electronics and Communication Engineering from Thapar Institute of Engineering and Technology, in Patiala, India, and Ph.D. from the National Institute of Technology, in Kurukshetra, India. He has conducted many workshops on Soft Computing and its applications in engineering, Wireless Networks, Simulators, etc. He has a keen interest in teaching and implementing the latest techniques related to wireless and mobile Communications. He has opened up a student chapter of IEEE as Branch Counselor. His research interests are routing protocol design, performance evaluation and optimization for wireless mesh networks using nature-inspired computing, Internet of Things and Space Communication, etc. He is a member of various professional bodies.

Dr. Shubham Tayal is working as an Associate Professor at Ashoka Institute of Engineering & Technology, in Hyderabad. He has more than 6 years of academic/research experience teaching at the UG and PG levels. He received his B.Tech (ECE) from Maharishi Dayanand University, in Rohtak, M.Tech (Electronics) from the YMCA University of Science & Technology, Faridabad, and Ph.D. from the National Institute of Technology, in Kurukshetra. He has published several papers in various international journals and conferences. He is on the editorial and reviewer panel of many international journals and books. He is a member of IEEE. He has received a Green ThinkerZ International Distinguished Young Researcher Award in 2020. His research areas include semiconductor devices, VLSI design, device-circuit co-design issues, machine learning, and IOT.

Prof. Swati Gupta has been working with the Panipat Institute of Engineering & Technology since Feb 2012. She received her Master's from NITTTR, Punjab University, in Chandigarh and is currently pursuing her Ph.D. from NIT, in Kurukshetra. She has more than 17 years of teaching and research experience. She has 20 research papers to her credit, 13 of which are published in different international journals and 7 in proceedings of different national conferences. She was a technical program committee member for the 3rd International Conference on Next Generation Computing Technologies. Her areas of interest include communication systems, wireless and mobile communication, digital system design, networking, and sensor networks.

Contributors

Ibrahim Muhammad Abdul
Nigerian Stored Products Research Institute
Kano, Nigeria

Abdulkadir Abdulkadir Adamu
College of Information Science and Engineering, Hunan University
Changsha, China

Khaldi Amine
Computer Science Department, Faculty of Sciences and Technology, Artificial Intelligence and Information Technology Laboratory (LINATI), University of Kasdi Merbah, 30000
Ouargla, Algeria

K. Aishwarya
ECE Department, Vardhaman College of Engineering
Hyderabad, India

Shafee Anwar
Department of Industrial and Systems Engineering, IIT Kharagpur
Kharagpur, India

Renu Bala
Research Scholar, Chaudhary Devi Lal University
Sirsa, Haryana

Rashmi Bhardwaj
Professor of Mathematics, University School of Basic & Applied Sciences (USBAS) and Head, Non-Linear Dynamics Research Lab, Guru Gobind Singh Indraprastha University
Delhi, India

Susheela Dahiya
School of Computer Science, University of Petroleum and Energy Studies
Dehradun, India

Souvik Das
Department of Industrial and Systems Engineering, IIT Kharagpur
Kharagpur, India

Sachin Dhawan
Ambedkar Institute of Advanced Communication Technologies and Research, Geeta Colony
New Delhi, India

R. Shashank Dinakar
ECE Department, Vardhaman College of Engineering
Hyderabad, India

Varsha Duhoon
USBAS, GGS Indraprastha University
Delhi, India

Kahlessenane Fares
Computer Science Department, Faculty of Sciences and Technology, Artificial Intelligence and Information Technology Laboratory (LINATI), University of Kasdi Merbah, 30000
Ouargla, Algeria

Monish Gupta
University Institute of Engineering and Technology, Kurukshetra University, Kurukshetra
Kurukshetra, India

Dr. Rashmi Gupta
NSUT, East Campus
New Delhi, India

Vishal Gupta
Bharat Electronics Limited
Bangalore, India

Tarun Jaiswal
Department of Computer Applications,
National Institute of Technology
(NIT)
Raipur (C.G.), India

Bhagyalaxmi Jena
Silicon Institute of Technology
Bhubaneswar, India

Keshav Kaushik
School of Computer Science, University
of Petroleum and Energy Studies
Dehradun, India

Mohammad Ayoub Khan
College of Computing and Information
Technology, University of Bisha
Bisha, Saudi Arabia

Dr Cheshta Jain Khare
Assistant professor
SGSITS
Indore, India

Dr Vikas Khare
Associate Professor, STME
NMIMS
Indore, India

A. Jaya Lakshmi
Assistant Professor, ECE Department,
Vardhaman College of Engineering
Hyderabad, India

J. Maiti
Department of Industrial and Systems
Engineering, IIT Kharagpur
Kharagpur, India

A. Mallaiah
Associate Professor, ECE department,
Gudlavallereu Engineering College
Gudlavalleru, India

Abhishek Mehta
Research Solar at Department of
Computer and Informative Science,
Sabarmati University, Ahmadabad,
Gujarat, India and Assistant Professor
in Parul Institute of Computer
Application, Parul University
Vadodara, Gujarat, India

Anita Mohanty
Silicon Institute of Technology
Bhubaneswar, India

Subrat Kumar Mohanty
College of Engineering Bhubaneswar
Bhubaneswar, India

Archita Nandi
Department of Research Innovation &
Consultancy, I. K. Gujral Punjab
Technical University
Kapurthala, Punjab, India

Moses Oluwafemi Onibonoje
Afe Babalola University
Ado-Ekiti, Nigeria

Manju Pandey
Department of Computer Applications,
National Institute of Technology (NIT)
Raipur (C.G.), India

G. Boopathi Raja
Velalar College of Engineering and
Technology
Erode, Tamil Nadu, India

A. Rajesh
ECE Department, Vardhaman College
of Engineering
Hyderabad, India

Arun Kumar Rana
Panipat Institute of Engineering and
Technology
Samalkha, Haryana, India

Sanjeev Kumar Rana
Department of Computer Science and
Engineering,
Maharishi Markandeshwar
(Deemed to be University)
Mullana, Haryana, India

Sumit Kumar Rana
Department of Computer Science and
Engineering,
Maharishi Markandeshwar
(Deemed to be University)
Mullana, Haryana, India

Trupti Rathod
Assistant Professor, Vidyabharti Trust
College of Master in Computer
Application
Bardoli, Gujarat, India

Ankur Ratmele
IET, DAVV
Indore, M.P., India

Kafi Redouane
Computer Science Department, Faculty
of Sciences and Technology,
Artificial Intelligence and
Information Technology
Laboratory (LINATI), University of
Kasdi Merbah, 30000
Ouargla, Algeria

Euschi Salah
Computer Science Department, Faculty
of Sciences and Technology Artificial
Intelligence and Information
Technology Laboratory (LINATI),
University of Kasdi Merbah, 30000
Ouargla, Algeria

Ayodeji Olalekan Salau
Department of Electrical/Electronics and
Computer Engineering, Afe Babalola
University, Ado-Ekiti
Ado-Ekiti, Nigeria

Rewa Sharma
Department of Computer Engineering,
J.C. Bose University of Science and
Technology, Faridabad
Haryana, India

Ramesh Thakur
IIPS, DAVV
Indore (M.P.), India

Priyanka Tripathi
Department of Computer Applications,
National Institute of Technology (NIT)
Raipur (C.G.), India

Dr H.K. Verma
Professor
SGSITS
Indore, India

Li Zhiyong
College of Information Science and
Engineering, Hunan University
Changsha, China

1 VANET-Based Intelligent Traffic Light Control System by Detecting Congestion Using Fuzzy C-Means Clustering Technique in a Smart City

Anita Mohanty[1,], Subrat Kumar Mohanty[2], and Bhagyalaxmi Jena[1]*
[1]Silicon Institute of Technology, Bhubaneswar
[2]College of Engineering Bhubaneswar, Bhubaneswar

1.1 INTRODUCTION

Year after year, the traffic density on roads has been rising worldwide due to rapid developments in the automotive industry and improvements in public living standards. This leads to the phenomenon called traffic jams. Traffic jams or traffic congestions have a direct impact on the economy [1] and the environment, and are stressful for motor vehicle drivers as they lead to a waste of time at a junction. To minimize these effects, there is need for developing a novel Intelligent Transportation System (ITS). Conventional traffic detection, control, and monitoring solutions are infrastructure-based. They include video and image processing, inductive loops [2,3], and microwave radars [4]. These methods are capable of providing traffic information of a fixed point or that of a shorter road section. They can detect traffic conditions at a particular location only and might not be able to provide information about traffic conditions over a large road segment. Thus, a Cooperative Vehicular Communication System [5] is considered an important ITS technology in a Vehicular Ad-hoc Network (VANET) to enhance the effectiveness of the system through continuous exchange of information between the vehicle and infrastructure (V2I Communication). VANET is a subtype of a mobile network with vehicles as nodes that run on a road and

DOI: 10.1201/9781003140443-1

1

communicate with one another as well as with infrastructure nodes through wireless technology [6–8]. In this technique, vehicles running on roads help gather information about other nearby vehicles and the possibility of a jam, and alternate the congestion conditions between them. In almost all detection techniques, including a gain amplifier [9], congestion recognition using wavelet technique [10], pattern recognition [11], and the number of messages collected from particular vehicles is large. Because of the limited availability of bandwidth in VANET, a clustering scheme is expected to be developed.

When the vehicles are in a traffic congestion, some of their parameters, such as speed, fuel consumption, and carbon dioxide (CO_2) emission, are very much similar. As these vehicles are in a dynamic state, the most popular clustering method i.e. FCM is preferred for detection of congestion. After detection of a congestion, the traffic jam is dispersed by changing the traffic light sequence as well as the duration of the green light at a junction.

Traffic lights are signaling devices installed at the junction of a road and are used to control the flow of traffic [12]. This chapter suggests an intelligent system to solve the question of bandwidth requirement for traffic information transmission, detect the congestion by using a clustering algorithm, and minimize the waiting time at junctions by reducing the queue. This can be accomplished by establishing traffic lights at junctions as controlling units. These lights alter the timing of its RED-YELLOW-GREEN cycle effectively depending on the traffic density in the lanes at a junction.

The rest of the chapter is organized as follows: Section 1.2 discusses the associated work on the simulation of a dynamic traffic light. The methodology to reduce the waiting time of vehicles at the intersections is provided in Section 1.3. The proposed method includes the extraction of vehicle parameters, the application of an FCM clustering technique to detect congestion, and LabVIEW for traffic light operation. Section 1.4 represents the overall model description. Section 1.5 describes the implementation of the method in detail. Section 1.6 represents the environment for simulation and the studies of the performance of the proposed scheme. Finally, we summarize the work proposed in this chapter in Section 1.7. Also, this section includes a short expression of future work.

1.2 LITERATURE REVIEW

VANET is a type of mobile network that considers vehicles as nodes that can interact with other vehicles and roadside units by transmitting and receiving messages. These vehicle nodes when facing a situation like traffic congestion due to fixed duration allotment to traffic lights at a junction, generate and transmit information to the roadside controller at that junction to control the sequence of light operation.

The traffic control systems help regulate the traffic flow at junctions, such as crossroads or zebra crossings [13]. At such junctions, traffic lights act as the essential control systems that manage the traffic flow [14]. This traffic control system should be able to manage the traffic flow density spontaneously by allowing vehicles to move in a regulated path [15,16].

The traffic regulation systems in Bhubaneswar, a city in India, are conventional systems with a fixed-time traffic light control arrangement [17,18].

Over the past two decades, researchers have proposed a number of congestion detection and control techniques to monitor, detect, and reduce congestion at the junctions within the city. Some authors have developed a variety of algorithms and controllers to dispose off the traffic flow at junctions [19]. Fuzzy logic controllers are designed to reduce the heavy traffic volume for pedestrians crossing the roads [20]. However, the duration of traffic lights is not adjusted by the intensity of traffic in a lane. To know the intensity of traffic in a lane, different techniques have been proposed. One author [21] proposed a technique to clear the traffic jam by examining the background features of lanes at a junction using many background images. This technique is real-time and also robust. Another author [22] developed a method to find out real-time traffic density from video images by analyzing digital data. However, during the estimation of density, if the direction of the digital camera is not similar for both the reference image and the actual image, there may be errors in the calculations. Thus, to minimize errors, the images must be taken in the same orientation. In the method suggested by Krauss et al. [23] and Cherrett et al. [24], inductive loops are fixed on the roads to identify vehicles passing over the loop. This works for congestion at all speeds, but has large errors in the detection of congestion and the dissemination of provided traffic information. The other disadvantages are that the installation of the inductive loop device is difficult and its maintenance is annoying. Inductive loops are a costeffective solution but have a high failure rate when used on the surface of a poorly constructed road. In Liepins and Severdak [25], the authors describe a method to count the number of vehicles using magnetic sensors in an intelligent traffic management system. These are not influenced by weather conditions; hence, when compared with other techniques, they provide more balanced and strong data for estimation. Yoon et al. [26] used GPS-enabled probe vehicles. Classified road networks broken into small segments are confined by traffic lights using traces of GPS collected from the probe vehicles. The vehicle speeds within each road segment are computed to determine the nature of traffic, whether it is congested or free-flowing. The authors proposed a clustering technique to identify the intensity of congestion. The vehicle nodes having similar characteristics are virtually grouped to form clusters using some procedures where each cluster will have one cluster center, which is responsible for inter-cluster conversation, local message aggregation, and local information propagation [27,28].

All these existing systems have their own solutions for congestion detection and traffic light control to disperse traffic. These existing congestion detection and control systems as proposed by the authors require special hardware devices, such as cameras, sensors, GPS, and others, which are expensive and there is a high chance of the sensors failing. The disadvantages of the above techniques include:

- Failure of an individual sensor
- Individual sensors covering a restricted location
- Affected by bad weather conditions and bad image capturing during the night
- Errors may arise if cameras are not properly aligned

The effective solution to the issues of these above individual techniques is a clustering technique in VANET and implementation in LabVIEW, where the vehicle data extraction is accurate, less affected by weather conditions, more reliable, and robust because it does not involve the use of any hardware.

The objective of this proposed work is

- To eliminate the failure of sensors for detection of congestion in a lane
- To improve the accuracy of congestion detection by using a Fuzzy C-Means clustering technique
- To reduce the need for hardware and network service charges
- To propose a method that is tested by implementing a prototype

1.3 PROPOSED METHOD

1.3.1 PARAMETER EXTRACTION

A traffic scenario was created using the Simulation of Urban Mobility (SUMO) simulator at the most crowded Kalinga Hospital Junction, Bhubaneswar. An FCM clustering algorithm was used to detect congestion, which works based on the parameters of vehicles extracted from the simulator. The process of the extraction of vehicle data using SUMO is shown in Figure 1.1.

http://openstreetmap.org was used to import the network of Kalinga Hospital Junction, which is shown in Figure 1.2(a).

The downloaded open street map was saved in an .OSM file format and carried to SUMO to generate traffic situations, which was saved as a .cfg file, as shown in Figure 1.2(b).

In the SUMO simulator, the congestion was created by delaying the duration of the green light of a traffic signal, which can be assumed as a delay in a green light in the lane where there is no vehicle, and in other lanes where there is congestion.

Then, for simulation purposes, the vehicles' raw data outputs, such as lane id, CO, CO_2, NO_x, PM_x, maximum speed, fuel consumption, noise, and others, were derived from the simulator.

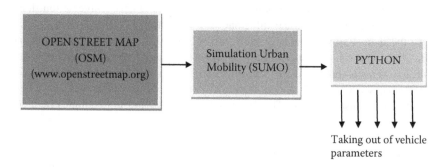

FIGURE 1.1 Process of extracting real-time parameters of vehicles.

(a) (b)

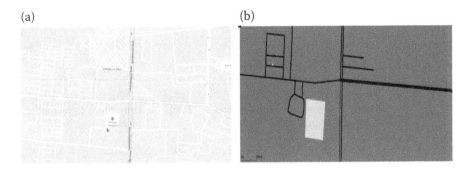

FIGURE 1.2 (a) Kalinga Hospital Junction, Bhubaneswar, original OpenStreetMap [20] and Figure 1.2. (b) The map imported from OSM in SUMO.

1.3.2 CLUSTERING ALGORITHM FOR CONGESTION DETECTION

Due to technological advancements in the automobile industry, new vehicles are being installed with various onboard sensors that detect vehicle speed (km/hr), CO2 emission (mg/sec), and fuel consumption (ml/s). As these messages are large in numbers, they sap the existing bandwidth in VANET. The vehicles' onboard control units convert these messages into a specific message using a clustering algorithm and transmit them to all vehicles. The FCM clustering algorithm [29,30] is preferred to work on the data set D for four lanes of the junction, as seen in Table 1.1 calculated using the algorithm in Appendix A, to categorize the messages with traces of speed, CO_2 emission, and fuel consumption, which are collected from the SUMO simulator. These data samples are gathered to form a number of clusters and then the distance between the centers of clusters is calculated to detect the congestion for each lane.

1.3.3 LabVIEW FOR TRAFFIC LIGHT OPERATION

The conventional systems for controlling traffic use a fixed time interval. These systems have a pretended cycle time. The present-day traffic light controllers (TLCs) make use of microprocessors and microcontrollers. These controllers have certain limitations because they use pretended hardware that works on the program and is not flexible with modifications, except by a programmer only. The waiting time for vehicles is longer and fuel consumption is high because of the fixed defined time intervals of green, yellow, and red lights. These controllers operate lights for a predefined duration and do not operate lights according to the density of vehicle traffic in a lane. Therefore, a lot of time is wasted by green lights on a low-congested road, compared to its counterpart. A Fuzzy Expert system and Artificial Neural Network are used by some adaptive traffic light controllers, but they are not user-friendly. Thus, to overcome these disadvantages, a traffic light control system using LabVIEW can be designed and implemented as it is a user-friendly Graphical Programming Language.

LabVIEW programs, called virtual instrumentations or VIs, use functions that shape the input received from the user interface or other sources and disclose that input [31].

TABLE 1.1

Data Set D

Number of Vehicles	LANE 1			LANE 2		
	Speed (Km/hr)	CO2 Emission (mg/sec)	Fuel Consumption (ml/sec)	Speed (Km/hr)	CO2 Emission (mg/sec)	Fuel Consumption (ml/sec)
1	0	2624.72	1.13	0	2624.72	1.13
2	1.63	3117.82	1.34	1.70	3180.87	1.37
3	4.26	4896.96	2.11	2.54	2743.50	1.18
4	27.17	8674.93	3.73	3.12	3655.93	1.57
5	27.37	5577.02	2.40	2.86	3460.67	1.49
6	9.20	7260.24	3.12	9.30	8382.62	3.60
7	6.10	5061.23	2.18	10.43	6983.71	3.00
8	3.01	3393.26	1.46	11.56	8971.36	3.86
9	1.52	3028.86	1.30	15.64	8475.07	3.64
10	27.44	5477.44	2.35	16.48	9001.59	3.87
11	8.49	7373.97	3.17	17.19	12536.11	5.39
12	4.98	4734.79	2.04	17.33	10371.44	4.46
13	2.44	2940.96	1.26	22.34	18006.88	7.74
14	21.28	5335.66	2.29	21.98	17680.25	7.60
15	10.72	8354.58	3.59	21.73	13322.30	5.73
16	26.52	4915.89	2.11	27.17	6020.83	2.59
17	4.89	5307.62	2.28	26.11	18035.66	7.75
18	21.78	6373.28	2.74	26.44	15044.55	6.47
19	26.79	6652.21	2.86	27.30	5297.24	2.28
20	22.21	8545.74	3.67	5.84	5584.19	2.40
21	6.65	5534.28	2.38	6.76	4407.81	1.89
22	7.97	5072.06	2.18	27.64	8549.24	3.67
23	6.54	6242.26	2.68	19.88	16330.06	7.02
24	27.51	10215.84	4.39	18.22	11147.94	4.79
25	8.95	4356.62	1.87	12.9	10553.88	4.54
26	6.92	2594.61	1.12	6.59	6385.81	2.74
27	7.04	5835.94	2.51	8.26	6578.98	2.83

Number of Vehicles	LANE 3			LANE 4		
	Speed (Km/hr)	CO2 emission (mg/sec)	Fuel Consumption (ml/sec)	Speed (Km/hr)	CO2 emission (mg/sec)	Fuel Consumption (ml/sec)
1	3.12	3655.93	1.57	5.08	4763.48	2.05
2	5.08	4763.48	2.05	6.71	4695.25	2.13
3	6.71	4695.25	2.13	9.30	8382.62	3.6
4	9.30	8382.62	3.6	11.56	8971.36	3.86

TABLE 1.1 (Continued)
Data Set D

Number of Vehicles	LANE 3			LANE 4		
	Speed (Km/hr)	CO2 emission (mg/sec)	Fuel Consumption (ml/sec)	Speed (Km/hr)	CO2 emission (mg/sec)	Fuel Consumption (ml/sec)
5	11.56	8971.36	3.86	4.25	5006.74	2.15
6	4.25	5006.74	2.15	14.10	11690.13	5.03
7	14.10	11690.13	5.03	6.18	5236.77	2.25
8	6.18	5236.77	2.25	15.54	8364.15	3.6
9	15.54	8364.15	3.6	8.26	6578.98	2.83
10	8.26	6578.98	2.83	17.94	14013.07	6.02
11	17.94	14013.07	6.02	20.00	13938.78	5.99
12	20.00	13938.78	5.99	10.97	5993.99	2.58
13	10.97	5993.99	2.58	21.73	13322.30	5.73
14	21.73	13322.30	5.73	12.64	7738.80	3.33
15	12.64	7738.80	3.33	23.09	11993.11	5.16
16	23.09	11993.11	5.16	15.00	11677.48	5.02
17	15.00	11677.48	5.02	24.40	12466.55	5.36
18	24.40	12466.55	5.36	26.63	20162.73	8.67
19	26.63	20162.73	8.67	19.51	14903.16	6.41
20	19.51	14903.16	6.41	27.39	10386.67	4.46
21	27.39	10386.67	4.46	24.13	17616.55	7.54
22	24.13	17616.55	7.54	29.23	15345.23	4.87

A VI has segments including a front panel, block diagram, connector pane, and icon. The user interface is the front panel. The block diagram contains a graphical source code, which outlines the function of the VI. The connector pane and icon allow the VI to be used in another VI.

This traffic light simulation model is designed using local variables, MATLAB, and flat Sequence Structure.

Local variables help access the objects of the front panel from several places in the block diagram of a VI. A flat sequence structure is an ordered set of frames that execute sequentially. It executes from one end to the other and until the last program is executed. The MATLAB code of the FCM algorithm is executed in the Math Script Node of LabVIEW.

1.4 MODEL DESCRIPTION

As the conventional system is not able to handle the uncontrolled traffic with varying density in metropolitan areas during office hours, a new adaptive light control must be implemented. To manage this varying traffic environment, the proposed system will be able to change traffic light sequence as per the traffic

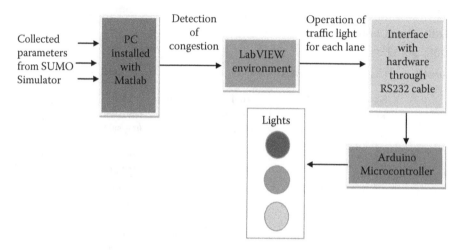

FIGURE 1.3 The block representation of the suggested adaptive traffic light operation control system.

intensity in a lane of a junction at any moment in a city. The block representation of the suggested model is displayed in Figure 1.3.

Here, the different parameters of the vehicles are input into the system. Then, the proposed model uses them in a clustering technique in the Math Script of LabVIEW to calculate the congestion in a lane at a junction, and further processes it to identify the green light status of the lane with red light in other lanes. The clustering algorithm forms a cluster by considering the vehicles with nearly similar parameters. Then, the distance between the centers of the clusters of different clusters is calculated to find the congestion, by comparing it with a predefined threshold value. When a congestion is detected in a lane, the sequence of the traffic light will change; starting from the congested lane with a green light and red light in other lanes. This will continue to operate in such a way until the traffic is dispersed or it will be altered again if another congestion is detected elsewhere.

1.5 SYSTEM IMPLEMENTATION

The overall implementation of the system is described in the following two sections.

1.5.1 SOFTWARE SECTION

The entire application program is developed using the LabVIEW graphical programming language. The basic functions used in the application development are Math Script, While Loop, and Flat Sequenced Structures.

FCM clustering algorithm is written in the Math Script of LabVIEW to detect congestion in a lane. For each lane of a junction, four scripts are written whose inputs are the text documents containing the parameters of the vehicles, which were extracted from the SUMO simulator. After the estimation of congestion in a lane at a junction, the next phase is to appropriately control the traffic signals.

(a) (b) (c)

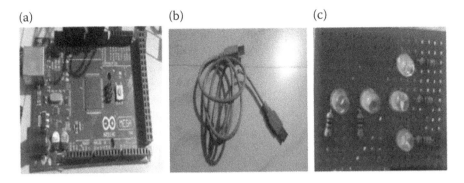

FIGURE 1.4 (a) Arduino Uno board, (b) USB cable, (c) LED and resistors.

1.5.2 HARDWARE SECTION

For implementation of the system prototype, the following hardware devices and components have been used (as shown in Figure 1.4): a USB cable, an Arduino microcontroller ATmega 2560, and LEDs. The USB cable is used to connect the PC with the LabVIEW output to the Arduino microcontroller for the operation of traffic lights (any specification). The ATmega 2560 has a 256 KB flash memory, 54 digital input/output pins, 16 analog inputs, 4 Universal Asynchronous Receiver/Transmitters (UARTs) ports, a 16 MHz crystal oscillator, a USB connection, a power jack, an ICSP header, and a reset button. Three different colors of LEDs are used with 1k ohm resistors to represent the traffic signal, whose operations are controlled by flat sequence and PC Port Number output.

1.6 EXPERIMENTAL RESULTS AND OBSERVATIONS

The efficiency of the FCM technique is tested using HP Intel (R), a Core (TM) i3-4000M CPU @ 2.40 GHz with a 4 GB RAM PC having MATLAB 2015 [32]. The simulation is performed in the LabVIEW environment. The data set with the parameters of vehicles is provided to the FCM algorithm for each lane, and the required number of clusters are created. The three-dimensional plot of the parameters for each of the vehicles in each lane is shown in Figure 1.5, where red X marks are the centers of clusters.

The Euclidian distance between the centers of the clusters in each lane is given in Tables 1.2, 1.3, 1.4, and 1.5. Out of these calculated distances, the distance between cluster heads of cluster-2 and cluster-3 in Lane1 is minimum. Which means there is congestion in Lane1.

Figures 1.6 and 1.7 show the graphical codes in frames of flat sequence structure to control the traffic signal starting from Lane1 where congestion is detected.

After the detection of congestion in Lane1, the output is utilized to manage traffic light in LabVIEW and, at the same time, the data is utilized in the Arduino

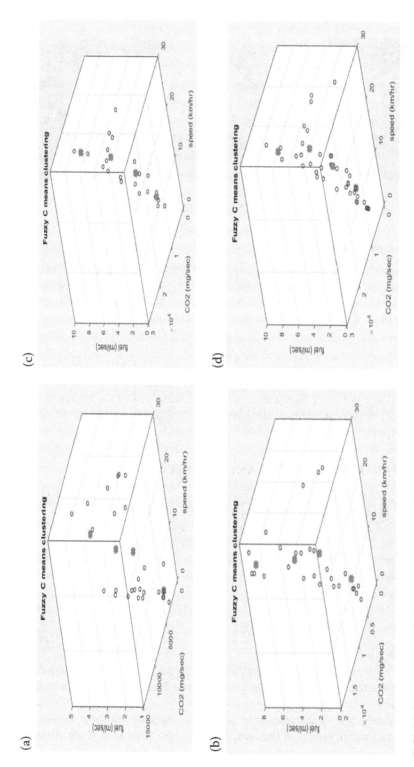

FIGURE 1.5 (a) 3-D plot of parameters for each of the vehicles with the center of clusters for Lane1, (b) 3-D plot of parameters for each of the vehicles with the center of clusters for Lane2, (c) 3-D plot of parameters for each of the vehicles with the center of clusters for Lane3, (d) 3-D plot of parameters for each of the vehicles with the center of clusters for Lane4.

TABLE 1.2

The Distance Between the Centres of the Clusters in Lane1 using FCM Technique

Centres of Clusters	Cluster-1	Cluster-2	Cluster-3	Cluster-4
Cluster-1	0	3653.7	2095.6	5869.7
Cluster-2	3653.7	0	1558.1	2216
Cluster-3	2095.6	1558.1	0	3774.1
Cluster-4	5869.7	2216	3774.1	0

TABLE 1.3

The Distance Between the Centres of the Clusters in Lane2 using FCM Technique

Centres of Clusters	Cluster-1	Cluster-2	Cluster-3	Cluster-4
Cluster-1	0	13616	4092	7862
Cluster-2	13616	0	9524	5754
Cluster-3	4092	9524	0	3770
Cluster-4	7862	5754	3770	0

TABLE 1.4

The Distance Between the Centres of the Clusters in Lane3 using FCM Technique

Centres of Clusters	Cluster-1	Cluster-2	Cluster-3	Cluster-4
Cluster-1	0	10344	13868	5911
Cluster-2	10344	0	3524	4433
Cluster-3	13868	3524	0	7958
Cluster-4	5911	4433	7958	0

microcontroller to manage the traffic lights mounted in the prototype. The LabVIEW program for the connection to Arduino and traffic lights is shown in Figure 1.7.

For the control of traffic lights, the Front Panel in LabVIEW is represented in Figure 1.8, which has the following modules:

TABLE 1.5

The Distance Between the Centres of the Clusters in Lane4 using the FCM Technique

Centres of clusters	Cluster-1	Cluster-2	Cluster-3	Cluster-4
Cluster-1	0	13480	5908	9770
Cluster-2	13480	0	7572	3710
Cluster-3	5908	7572	0	3862
Cluster-4	9770	3710	3862	0

- Indicators to display the count of vehicles in each lane
- For each lane, 3xLEDs indicators, representing the traffic lights

The duration of the green light depends on the range in which the minimum distance between the clusters for a lane falls. If the minimum distance is less, the duration of the green light is more. ATmega 2560 microcontroller, LEDs, and resistors are used to implement the prototype to fulfill the requirement of a traffic light control system. The experimental setup is shown in Figure 1.9.

After the implementation of the prototype and compilation, the LabVIEW program is uploaded to the Arduino board and the LEDs glow in a sequence.

This proposed method is compared with the regular method on different aspects. The different aspects and their comparisons are listed in Table 1.6.

From this comparison, it is clear that the proposed method is better in all aspects than the regular method.

1.7   CONCLUSION

The fixed duration traffic light is a very critical issue that affects both the residential and governmental personnel in a city. Recently used conventional traffic systems, which are low in efficiency, affect the economic, health, finance, and environmental fields. This proposed prototype tries to give society an improvised traffic congestion detection and traffic light control system. It consists of a controller that manages the lights of a junction of mono-directional roads. Based on the information about detected congestion, the time decided for the green light will be increased to allow for a larger flow of vehicles in case of a jam, or decreased to avoid unnecessary waiting time when there are no vehicles in other lanes. The designed prototype is implemented, realized, and verified to ensure complete validation of its operations and functions. In the future, some improvements can be added in the form of a button for pedestrians crossing roads, a display of delay timing, as well as car accident and failure modes. The simultaneous operation of different traffic controllers at several junctions will be checked in the future to achieve complete synchronization.

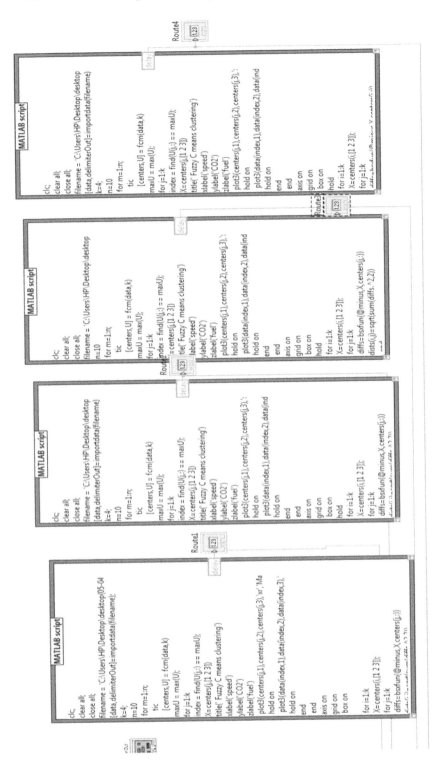

FIGURE 1.6 The proposed system block diagram (LabVIEW) to detect congestion.

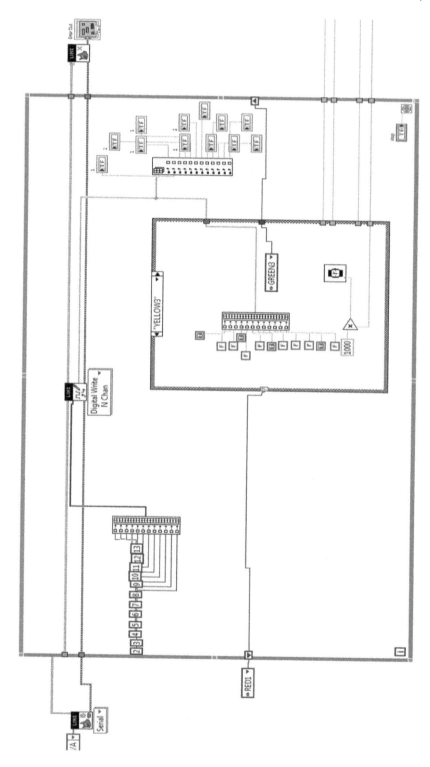

FIGURE 1.7 The proposed system block diagram (LabVIEW) to connect Arduino to operate Traffic Light.

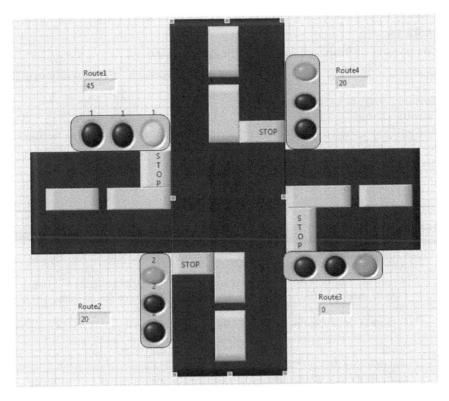

FIGURE 1.8 LabVIEW Front panel for detecting congestion and controlling traffic lights.

FIGURE 1.9 Experimental setup.

TABLE 1.6

Comparison of Proposed Method with Regular Method

Aspects	Regular Method	Proposed Method
Prerequisite Traffic personnel in peak hours	Yes	No
Congestion chances	High	Low
Chances of Varying duration of traffic light	No	As per requirement
Waiting time in traffic	More	Less
CO_2 emission	More	Less

REFERENCES

1. Schrenk D., Lomax T. "The 2009 annual mobility report," Texas Transp. Inst., Texas A&M Univ., College Station, TX, Tech. Rep., 2009.
2. Wright M., Horowitz Roberto."Fusing loop and GPS probe measurements to estimate freeway density," IEEE Transactions on Intelligent Transportation Systems, vol. 17, no.12, pp. 3577–3590, 2016.
3. Guibert D. *et al.* "Robust blind deconvolution process for vehicle reidentification by an inductive loop detector," *IEEE Sensors Journal*, vol. 14, no.12, pp. 4315–4322, 2014.
4. Mandalk Sen A *et al.* "Road traffic congestion monitoring and measurement using active RFID and GSM technology," Proceedings of International IEEE Conference on Intelligent Transportation Systems, pp. 1375–1379, 2011.
5. Ramon Bauza, Gozalvez Javier, Sanchez-Soriano Joaquin "Road traffic congestion detection through cooperative vehicle-to-vehicle communications," IEEE Local computer Network Conference, pp. 606–612, 2010.
6. Ahmed H., Pierre S., Quintero A. "A flexible testbed architecture for VANET," *Vehicular Communication*, vol. 9, pp. 15–126, 2017.
7. Yoo H., Kim D. "Repetition-based cooperative broad casting for vehicular ad-hoc networks," *Computer Communication,* vol. 34, no. 5, pp. 1870–1882, 2011.
8. Fußler H., Mauve M., Hartenstein H., Kasemann M., Vollmer D. "Location-based routing for vehicular ad-hoc networks," *Mobile Computing and Communications Review*, SIGMOBILE, New York, 2002.
9. Wei Q., Yang B. "Adaptable vehicle detection and speed estimation for changeable urban traffic with anisotropic magneto resistive sensors," *IEEE Sensors Journal,* vol. 17, no. 7, pp.2021–2028, 2017.
10. Sen R., Siriah P., Raman B. "RoadSoundSense: A coustic sensing based road congestion monitoring in developing regions," Sensor, Mesh and Ad Hoc Communications and Networks, SECON'2011, Salt Lake City, Utah, USA, 2011: 125–133.
11. Roy A., Gale N., Hong L. "Automated traffic surveillance using fusion of Doppler radar and video information," *Mathematical and Computer Modelling*, vol. 54, 531–543, 2011.
12. Kham Nang Hom, New Chaw Myat. "Implementation of modern traffic light control system," *International Journal of Scientific and Research Publications*, vol. 4, no. 6, pp. 14–20, June 2014.
13. Megasari R., Lukman, Novianingsih K. "Traffic light delay optimization at the intersection with fuzzy logic mamdani method," *Jurnal EurekaMatika,* vol. 5, pp. 10–19, 2017.

14. Kaushal N., Joshi V. "Fuzzy logic controller based traffic light system," *International Journal of Advanced Research in Computer Science and Software Engineering*, vol. 6, pp. 223–229, 2016.

15. Sutomo B. "Modelling of traffic light control system based on vehicle density with edge detection and fuzzy logic techniques," *Informatika*, vol.15, pp. 116–126, 2015.

16. Salau A. O., Yesufu T. K. "A probabilistic approach to time allocation for intersecting traffic routes," *Advances in Intelligent Systems and Computing,* vol. 1124, 2020. Springer, Singapore. https://doi.org/10.1007/978-981-15-2740-1_11

17. Homaei H., Hejazi S. R., Dehghan S. A. M. "A new traffic light controller using fuzzy logic for a full single junction involving emergency vehicle preemption," *Journal of Uncertain Systems*, vol. 9, pp. 49–61, 2013.

18. Nugroho E. A. "Traffic light controller system based on fuzzy logic," *J. SIMETRIS*, vol. 8, pp. 75–84, 2017.

19. Chiu, S. "Adaptive Traffic Signal Control Using Fuzzy Logic," Proceedings of the IEEE Intelligent Vehicles Symposium, 1992, pp. 98–107.

20. Niittymaki J., Pursula M. "Signal Control Using Fuzzy Logic. Fuzzy Sets and Systems," vol. 116, pp. 11–22, 2000.

21. Xie Lei, Qing Wu, Xiumin Chu, Jun Wang, Ping Cao. "Traffic jam detection based on corner feature of background scene in video based ITS," IEEE International Conference on Networking, Sensing and Control, ICNSC 2008, pp. 614–619, 2008.

22. Ruddaba Awan, Rani F. "Video based effective density measurement for wireless trafffc control application," International Conference on Emerging Technologies, pp. 99–101, 2007.

23. Krause B., von Altrock C., Pozybill M. "Intelligent highway by Fuzzy Logic: Congestion detection and traffic control on multi-lane roads with variable road signs," Proceedings of EUFIT`96, Aachen, Germany, 1996.

24. Cherrett T., Waterson B., McDonald M. "Remote automatic incident detection using inductive loops," Proceedings of the Institution of Civil Engineers: Transport, pp. 149–155, 2005.

25. Liepins M., Severdaks A. "Vehicle detection using non-invasive magnetic wireless sensor network," IEEE Publication Year: 2013, pp. 601–604, 2013.

26. Yoon J., Noble B., Liu M. "Surface street traffic estimation," International Conference on Mobile Systems, Applications and Services-MobiSys'07, pp. 220–232, 2020.

27. Toor Y., P. Muhlethaler, A. Laouiti, A. D. L. Fortelle, "Vehicle ad hoc networks: applications and related technical issues," *IEEE Commun. Surveys & Tutorials*, vol. 10, no.3, pp. 74–88, 2008.

28. https://www.openstreetmap.org/#map=16/20.3147/85.8161

29. Mohanty A., Mahapatra S., Bhanja U. "Traffic congestion detection in a city using clustering techniques in VANETs," *Indonesian Journal of Electrical Engineering and computer Science,* vol. 5, no.3, pp. 401–408, March 2017.

30. Chang Chih-Tang, Lai Jim Z. C., Mu-Der Jeng. "A Fuzzy K-means clustering algorithm using cluster center displacement," *Journal of information Science and Engineering,* vol. 27, pp. 995–1009, 2011.

31. Dhiman A., Kumar A., Kumar A., Sokhal Y., Kaur Channi H. "Modeling of traffic light control using LABview," *International Journal of Scientific Research in Computer Science, Engineering and Information Technology*, vol. 2, no.6, pp. 539–542, 2017.

32. Math works http://www.mathworks.com

33. Ghosh S., Dubey S. K. "Comparative analysis of K-Means and Fuzzy C-Means algorithms," *International Journal of Advanced Computer Science and Applications,*vol. 4, no. 4, pp. 35–39, 2013.

APPENDIX B

Algorithm: Fuzzy C-means clustering [33]
Input:

- k: the number of clusters
- X: the data set containing m objects
- f: the parameter in objective function
- ε : a threshold for the convergence criteria

Output:

- A set of k clusters

Method:
Initialize $V = \{v_1 , v_2, ...v_k\}$
$V^{previous} \leftarrow V$
Compute the membership function $\mu_{C_i}(x)$ using Step (2).

Update the prototype v_i in V using Step (3) till $\sum_{i=1}^{C} \|v_i^{previous} - v_i \leq \varepsilon\|$; otherwise, return to Step (2).

---

2 Internet of Things Advancements in Healthcare

Keshav Kaushik[1], Dr. Susheela Dahiya[1], and Dr. Rewa Sharma[2]
[1]School of Computer Science, University of Petroleum and Energy Studies, Dehradun, India
[2]Department of Computer Engineering, J.C. Bose University of Science and Technology Faridabad, Haryana, India

2.1 INTRODUCTION TO IOT IN HEALTHCARE

Nowadays, physical devices with data storage and communication capabilities are connected by a global infrastructure, which allows identification, sensing, and connection, serving as a foundation for multiple services and applications. This entire global infrastructure is referred to as the "Internet of Things (IoT)". IoT finds a wide range of applications in the healthcare sector, in that it provides flexibility in connecting sensors to mobile devices, mobile devices to humans, doctors to machines, patients to machines, and tags to readers. In fact, the entire IoT infrastructure connects humans, smart devices, machines, and intelligent systems, thereby increasing the overall effectiveness of the healthcare system. However, growing adoption of IoT technology in various fields of healthcare has led to several concerns with regard to privacy and security, which are increasing at an alarming rate. Privacy and security both are distinct terms; however, they are synonymously used with regard to confidentiality. IoT comprises a general architecture and well-defined components, which, together, make use of various technologies to help perform several tasks. However, these components are also associated with privacy and security issues, which need to be handled carefully. The IoT is progressively gaining ground as an effective standard in recent wireless communications. This is because of the latest advancements in embedded systems, communication technologies, and standards [1]. It is now being considered as a prominent way of integrating the latest communication and computing standards. IoT allows various uniquely identified devices to be interconnected via a network and to the global Internet ubiquitously. It provides us the ability to feel, perceive, and respond to surroundings through integration of various technologies, such as Radio Frequency Identification (RFID), cloud computing, real-time location, and embedded sensors. IoT, uniquely and digitally, recognizes all the devices interconnected via IoT

DOI: 10.1201/9781003140443-2

networks and provides addressing schemes for synchronized functioning to achieve certain common objectives.

The IoT is not a single technology; it is a combination of various technologies that together make things simpler and easier e.g. communication technology, actuator technology, electronic sensor information technology, as well as some of the recent advancements in analytics and computing. The combination of these technologies has given rise to various privacy and security issues in the IoT domain. In the healthcare sector, privacy plays a crucial role in safeguarding patient data, such as medical history and current conditions, that are to be kept confidential. Any privacy and security-related issues arising due to the use of IoT in healthcare applications must be prevented. However, due to advancements in IoT technologies, cases of cyber threats are increasing rapidly, which need to be handled proactively. The chapter discusses the role of IoT in the healthcare industry by explaining the entire IoT architecture, including its layers, components, and communication technologies. IoT protocols and safety measures, as well as IoT-based healthcare applications, are also highlighted in this chapter. The chapter also enlightens readers about the growth of IoT in a changing data environment. It explores the future aspects of IoT in healthcare as well as the scope of telehealth and self-monitoring mobile apps. For instance, wearable IoT devices and IoT applications are proving to be beneficial for IoT-based healthcare applications developers, medical trainers, technology enthusiasts, specialists, software developers, scientists, PhD scholars, and researchers.

2.1.1   THE INTERNET OF THINGS AND HEALTHCARE

The IoT has revolutionized the existing technology as it is helping create and build smart infrastructure like smart cities, smart industries, smart agriculture, smart transportation systems, smart energy, and smart healthcare. It is anticipated that the coming years will observe a transformation in the way healthcare is provided. Traditional healthcare systems focus more on offering a hospital-home balanced healthcare system, while IoT is expected to provide more home-centered care systems [2]. As observed during the COVID-19 pandemic, there is need for transformation in the current healthcare systems. Due to the rapidly increasing cost of healthcare and rising prevalence of chronic diseases, a necessary paradigm shift in the healthcare system is required i.e. from hospital-centered systems to an individual-centered environment, with more focus on providing optimum care to patients [3]. This much-needed change demands the integration and overlap of various IoT technologies and architectures for building smart healthcare domains [3–6]. Privacy and security are the major concerns, as the devices communicate wirelessly [7]. Security is a serious concern where IoT networks are deployed at large scales. IoT-based healthcare systems collect data from various wearable and non-wearable devices that communicate through a wireless network. Therefore increasing the risk of security and privacy breaches [8]. It is very important to ensure secure, reliable, and robust exchange of information between devices, patients, and caretakers. Preventing the misuse of data and ensuring reliable and secure communication will boost the adoption of IoT-based healthcare applications.

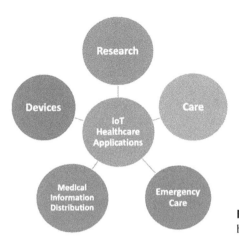

FIGURE 2.1 Applications of IoT in healthcare.

It is not feasible to reutilize the traditional security and protection techniques that are already a part of the existing cryptographic mechanisms and reliable protocols in IoT networks due to constrained resources, system architecture, and high-end security mechanisms [9].

Figure 2.1 displays the various applications of IoT in the healthcare domain. There are a number of applications of IoT in healthcare, including research, wearable and non-wearable devices, care, medical information distribution, and emergency care.

Healthcare applications based on IoT are different than those functioning on current wireless networks, therefore the need for security is much higher. Several crucial issues are required to be addressed. Efficient security solutions must be provided with regard to limited memory, processing power, and communication bandwidth. Thus, it is not feasible to use traditional encryption methods. Also, medical sensors used are quite small in size, therefore, they can easily be seized or lost. Recently, several attempts have been made to design a Smart e-Health Gateway for health applications based on IoT [3]. A gateway acts as a bridge between the Internet and a sensor network. These gateways are stationary and are not resource-constrained in terms of memory, power consumption, or communication bandwidth. Therefore, gateways can assist in reducing the load of medical actuators and the number of sensors being used, by responsibly performing certain tasks.

Thus, smart gateways are capable enough to deal with a large number of issues such as reliability, scalability, and security, which are faced by pervasive healthcare systems.

2.1.2 THE GROWTH OF THE INTERNET OF THINGS IN A CHANGING DATA ENVIRONMENT

An increase in the amount of IoT data being generated impacts the technological landscape, affecting both information technology and operational technology. Large IoT data has a significant impact on IT infrastructure and technological standards.

The IoT provides us with a wide range of connectivity services and solutions that help us analyze and process the generated data e.g. edge computing. IoT also offers us hardware devices, such as gateways, sensors, actuators, and all other smart devices and objects. Moreover, it also provides various IoT platforms. All of these leverage the large data generated by IoT devices and work in sync to accomplish the common objective of transforming knowledge and insights gained from this data into some intelligent decisions and outcomes.

Although the number of big manufacturing operations and IoT projects is increasing every year, the IoT data collected by companies in these projects is not fully explored. It has been observed that companies utilize a small percentage of data and the remaining unexplored data is simply stored, which is termed as "dark data." A survey reported a case where only 1% of the data was analyzed out of the total data produced by 30,000 sensors deployed in an oil rig, and the remaining 99% of the gathered data was not explored at all, and even failed to reach operational decision-makers. This poses a significant challenge in fully exploiting the huge amount of data generated and collected by IoT devices to optimize the process and predict even more accurately.

Big Data includes information from several heterogeneous sources; therefore, storing, managing, and analyzing Big Data is quite difficult. Various heterogeneous sources of data make use of different techniques and ways of collecting data, such as through automated reports, various sensors, and by analyzing historical trends, and more. Different data sources may lead to conflicting information. For ensuring data integrity and quality throughout the process, it is crucial to determine false, incorrect, and redundant data, at the very beginning of the process.

Different file architectures are used by different systems. Older systems used File Allocation Table (FAT) architectures, whereas personal computers nowadays use New Technology File Systems (NTFSs). However, these architectures are not suitable for processing huge datasets generated by IoT networks and other such applications. Professionals use platforms like Xively, Carriots, Plotly, AMEE, Axeda, Thingspeak, Connectrra, and Exosite for storing, processing, and managing Big Data. Some popular operating systems supporting IoT devices and operations are Windows 10 for IoT, RIOT OS, and Google Brillo. Currently, data management is underway with its phenomenal growth and development. For effectively and efficiently handling real-time data, it is important to choose the right architecture and operating system.

2.1.3 IoT ANALYTICS IN HEALTHCARE

Healthcare can be defined as the maintenance and improvement of an individual's health by providing health-related services, such as proper preventive measures, diagnosis, and treatment of disorders or diseases. Healthcare services aim to achieve efficiency, effectiveness, patient safety, and timeliness. Access to healthcare may vary across individuals, communities, and nations, depending on health policies, social conditions, economic state, and many other factors. The healthcare industry is important for the well-being of the population around the world. With the explosive growth in global population, the demand for a robust healthcare industry has increased. Increase in incidence of critical illnesses has led to high demand for more resources, such as improved hospital infrastructures, medical experts, and others. IoT-based healthcare systems can help provide high-quality and intensive care to critical patients by using the latest technologies, thereby also assisting traditional healthcare systems. IoT-based healthcare services are apt for providing remote health monitoring at an affordable cost. This can lead to an improved economy, a nation's empowerment, and rapid industrialization.

Several concerns in the healthcare domain need attention, such as minimizing communication delays in emergency cases, an automated compilation of patient information, and providing remote healthcare services in deprived regions. IoT-based healthcare systems hold significant potential in addressing the above-mentioned issues efficiently.

Figure 2.2 shows an IoT-based architecture for healthcare monitoring. Various sensors are integrated with IoT devices to keep track of an individual's health and information. These sensors include wearable sensors, temperature sensors, climatic sensors, and others. For example, smartwatches or fitness bands keep track of body temperature, heart rate, blood pressure, GPS location, and much more. Sensors that are attached to clothing determine pressure rate, EMG, and ECG of the patient. The recorded data is sent to the cloud for storage and analysis purposes. An appropriate message is conveyed to clients based on the recordings made.

Patients in intensive care units (ICU) need 24-hour attention, which is usually done by nurses. Nevertheless, this is quite a tedious task, as minor negligence may seriously impact a patient's health. Prajapati *et al.* [10] proposed a real-time system for regular and continuous monitoring of patients in ICU. A monitor placed bedside of the patient displays blood pressure, ECG, respiration, cardiac, and standard diagnosis. This monitor continuously records and sends patient's health parameters, such as pressure in the patient's brain, blood pressure, oxygen level, and heart rate, to the server through the internetwork. When these parameters reach above the threshold limit, a quick notification is sent to the respective doctor as well as to the ICU, thereby allowing early precautionary intervention. This real-time architecture is designed to reduce the probability of human errors, accelerate communication, and allow continuous monitoring of patients.

Baali *et al.* [11] designed a wearable power-aware device for providing real-time healthcare assistance to elderly people. Sensors interconnected with other devices over IoT networks are used to provide Technology Enabled Care (TEC) by measuring and monitoring various health parameters of the patient.

Figure 2.3 depicts the significance of Big Data Architecture in the healthcare domain. Big Data Analytics [12] has significantly proven its advantages in improving the quality of life of individuals by helping in predicting diseases, avoiding preventable deaths, and curing critical illness. The system gathers and integrates structured or unstructured data from various sensors. This data is processed and visualized using cloud infrastructure. Insights gained through analysis help the system make intelligent decisions and a suitable message is sent to the relevant individual, who may be a remote person, doctor, hospital, or an ambulance service.

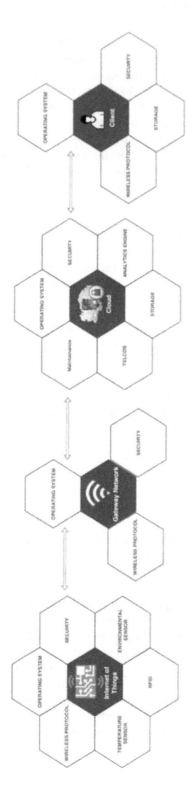

FIGURE 2.2 IoT-based architecture for monitoring health.

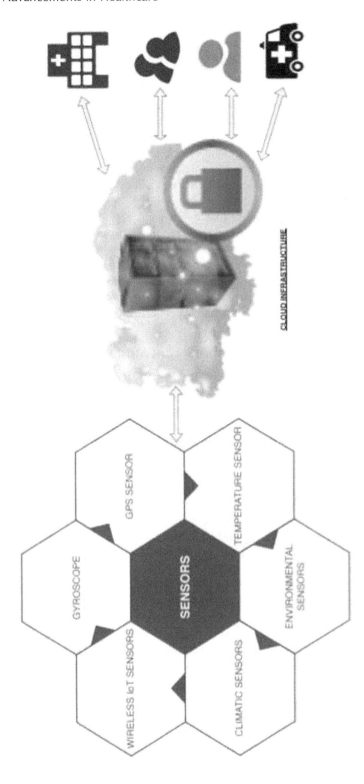

FIGURE 2.3 Significance of Big Data Analytics in healthcare domain.

2.1.4   FUTURE ASPECTS OF IoT IN HEALTHCARE

IoT devices and sensors offer viable solutions for various healthcare applications. Elderly patients and adults can enhance their quality of life by receiving better insights into changes in their health conditions using the data obtained through various wearable devices [13]. Security is a major concern in IoT-enabled healthcare systems. Data related to a patient's health is transferred over wireless network and, thus, might be intercepted by malicious users. Such kind of security and privacy threats may pose significant risk to patients' lives. Future technologies and architectures should therefore be aimed at providing secure and reliable real-time healthcare applications.

IoT data related to health generated by sensors and devices need proper authentication provided by the server to prevent malicious and unauthorized access. Networks should be able to change topology dynamically to support a secure and reliable transfer when the patient being monitored moves from one network to other. Future IoT healthcare systems aim to achieve high-throughput, effective sensing, reduced computation, and communication cost. The latest telecommunication technologies can help in minimizing time synchronization delay experienced between base stations and sensors. Smart healthcare systems must be able to extract and integrate contextual information along with data collected through IoT sensors [14].

2.1.5   TELEHEALTH AND SELF-MONITORING WITH MOBILE APPLICATIONS

Rapid globalization across the world has led to an increase in demand for critical care management by a population deprived of medical facilities. Hyang *et al.* [15] investigated the usage trends of telehealth counseling among Koreans suffering from high blood pressure. The authors also assessed the impact of self-monitoring mobile applications on the blood pressure of patients. Results showed a noticeable difference in the average change of systolic blood pressure at three months among the group that used self-monitoring applications and the group that did not use any such applications.

Conventional health services require people to physically visit the clinic to obtain appropriate care. The e-health system promotes collaboration between doctors and patients through information and communication technology. On the other hand, the telehealth system remotely tracks patients using cellular equipment. The Smart Health Idea is an expansion and enhancement to the e-health and telehealth frameworks. Real-time sensors track patients, provide diagnosis, and give fast updates to physicians. The definition of smart health has been applied by Rani *et al.* [16]. The authors developed a smart health architecture, in which the patient signs in with a smartphone application with a unique ID. All the patient information is stored under the respective ID and can be accessed anywhere and anytime through the application. The information is updated regularly and can be refreshed periodically to obtain latest data.

2.2   WEARABLE SENSORS AND IoT APPLICATIONS IN HEALTHCARE

A sensor is an electronic device or a module that is designed to detect specific events or changes in its environment and transmit the specified data to other electronics, which could be a computer processor or any other IoT device. An important aspect of the sensors is a measure of their sensitivity. A good sensor is supposed to follow certain rules, as enlisted below:

- It is sensitive to the property being measured
- It is insensitive to the other properties that it might encounter
- It does not influence or alter the property it is measuring

Another important characteristic of a sensor is its resolution. The resolution of a sensor is simply the smallest amount of change that it can detect, before giving the desired output.

Sensors are broadly classified into two categories, active and passive. More features of the smart sensors could include:

- Early warning for medical issues
- Neural technologies
- Automated smart medical systems
- Chronic condition management

In the world of IoT, there are numerous types of sensors being used today. We will focus mainly on healthcare-related IoT sensors. The sensors used in the healthcare sector are used to constantly monitor vital signs of patients and either store the data or transmit it wirelessly to healthcare professionals. The various types of sensors used today include:

- Smart wearables such as digital watches and fitness bands, which are embedded or attached to clothes and accessories. These typically contain sensors for motion, heart rate, and temperature.
- Ingestible pills with embedded sensors that can be dissolved inside the body.

Typically, biosensors, temperature sensors, and flow analyzer sensors are embedded.

- Blood sampling sensors, such as glucose measurement devices – these have blood analyzer (chemical) sensors.
- External sensors, such as pulse oximeter and blood pressure cuffs typically use pressure, heart rate, temperature, oxygen, and level (chemical) sensors. In addition, external sensors such as image sensors are used in radiography, cardiology, dental imaging, endoscopy, and other sectors.
- Epidermal sensors that are implantable, such as digital tattoos and patches. These include temperature, motion, heart rate sensors, and more.
- Tissue embedded sensors inside the body, such as pacemakers & defibrillators – these not only have sensors including accelerometer, heart rate sensors, and temperature sensors, but they also have actuators for performing certain functions when required.

2.2.1 IoT Devices and Protocols

One of the main features and requirements of any wearable IoT device is being able to communicate, either with another smart device or directly over the network or Internet. A protocol defines the set of rules with the help of which the IoT devices communicate with each other. There are a number of protocols being used in the world of networking, and different protocols have different advantages and disadvantages for a specific type of communication between devices or networks. Some of the important protocols used between wearables and IoT devices are as follows:

- **Bluetooth Classic**
  Also known as Bluetooth High speed, this is a preferred solution for transferring high bandwidth data, like audio streaming. It also consumes a lot of energy. Bluetooth protocols include Headset Protocol (HSP), Object Exchange (OBEX), Video Distribution Protocol (VDP), Audio Distribution (A2DP), and File Transfer Protocol (FTP).
- **ZigBee**
  ZigBee is a wireless network mesh protocol intended to transmit tiny data packets over small periods while using relatively low power. A ZigBee uses IEEE (Institute of Electronics and Electronics Engineering) 802.15.4 protocol, with a worldwide 2.4 GHz ISM frequency range, meaning it can be used almost everywhere. ZigBee protocols are mainly used for home automation, surveillance system, smart lighting, etc.
- **Z-Wave**
  Z-Wave is similar to ZigBee, as it is also a wireless mesh network technology, which uses low power, carries small amounts of data over medium distances. Both technologies use the same IEEE 802.15.4 standard and are widely used in local area sensor networks.
- **Wi-Fi**
  Wi-Fi is a Local Area Network (LAN) designed to provide Internet access wirelessly over a limited range. The structure of Wi-Fi is a start network, with a central hub, and all nodes or devices connect to it from around the hub. Its range is typically around 30 to 200 meters.
- **BLE (Bluetooth Low Energy)**
  The Bluetooth Special Interest Groups (SIGs) introduced this protocol. The main advantages are:
  - **Low power usage-** Compared to Bluetooth Classic, the power consumption of BLE is between 0.01% to 0.5% only.
  - **The smaller size of electronics-** The dimensions of the smallest Bluetooth Classic modules are 11 mm x 13 mm x 2 mm.
  - **Cost factor-** A BLE module would typically cost anywhere between USD 0.1 to 4, as compared to USD 0.5–6 for the Bluetooth Classic.
    To maintain a high standard of security, various Machine-2-Machine (M2M) protocols are being researched and used in IoT systems, in

wearables, or other devices that incorporate sensors and transmit data regularly. Some of these protocols are as follows:

- **REST**
  Representational State Transfer protocols use the standard HTTP (HyperText Transfer Protocol) application protocol, built on top of the TCP, UDP, and SSDP layers, for data transfer.
- **CoAP**
  Constrained Application Protocols (CoAP) is a lightweight protocol built on the UDP layer, designed for IoT M2M communication with low overhead. CoAP is designed for simple electronic devices that make use of 8-bit microcontrollers, low memory, and a TCP/IP communication stack.
- **MQTT**
  MQ Telemetry Transport is a lightweight application protocol, which is built on top of the layer. It is the OASIS standard for M2M communications. MQTT is a publish and subscribe messaging protocol, based on three components:
  - **MQTT broker (server):** Is capable of transmitting messages received from a collection of MQTT writers to a set of MQTT subscribers.
  - **MQTT Publishers:** Are liable for posting the correct message or material on the right subjects.
  - **MQTT Subscribers:** Accept all messages written on subjects on which they have been subscribed.

- **MQTT-SN**
  MQTT for Sensor Networks (MQTT-SN) is an extension of MQTT specifically designed for embedded sensors and non-TCP/IP wireless network (e.g. ZigBee), which processes specific constraints like low bandwidth, short messages, limited hardware and energy resources, and unreliable connections.
  - **MQTT-SN Clients:** They can interface with the traditional remote MQTT Broker via the MQTT-SN Gateway and use the MQTT-SN communication protocols.
  - **MQTT-SN Gateways:** Intermediate between MQTT-SN Clients.
  - **MQTT-SN forwarders:** Are used for correspondence with MQTT-SN clients while the MQTT-SN portal is not immediately available.

- **AMQP**

Advanced Message Queuing Protocols (AMQP) is an OASIS open-source framework protocol, developed at the top of the TCP layer. This is ideal for a message-oriented middleware and a distributed framework. AMQP is currently being considered as an interesting communications solution in cloud computing, M2M, and IoT applications. This protocol can be used on any computer and network and in any language.

2.2.2 IoT Applications in Healthcare

With advancements in IoT, healthcare is now more intelligent and is able to efficiently collect more health-related information than before.

Some of the major advantages of IoT applications in healthcare are:

- **Healthcare cost-saving:** Remote care is made possible by connected wearable devices. A physical meeting is also not required when healthcare information is already in the hospitals' databases. This may lead to low costs in healthcare; doctors can send alerts and reminders to patients by looking at the information available in the hospital's database.
- **Medical inaccuracy minimization:** Many people do not prefer visiting a doctor and instead self treat using information available on the Internet, but IoT healthcare devices help solve this problem of medical inaccuracies, because they provide more accurate results on a real-time basis; thus getting diagnosed through IoT devices is more reliable and safe.
- **Enhanced patient-healthcare experience:** Health progress can also be monitored with the help of IoT applications that provide a more realistic and enhanced experience to patients.
- **Effective chronic illness management:** Continuous health checkups and diagnoses are the epitomai of chronic care management. IoT has provided more advanced scopes for monitoring patients remotely.
- **Patient healthcare awareness:** Nowadays, most of the people wear healthcare fitness devices and many households have blood sugar and pressure monitors. All these advanced devices have created an awareness among people using them, thereby helping them take care and improve health.
- **Helping individuals with disabilities**: There are many IoT-enabled healthcare devices available in the market that help individuals with disabilities. Some of these devices are smart connected eyeglasses, artificial intelligence apps that can read surroundings, smart wheelchairs, and adjusting and self-learning hearing devices.

Some of the healthcare and wellness monitoring devices help in the early detection of diseases, healthcare surveillance, monitoring at home, and augmented reality in healthcare. All these IoT applications result in cost reduction through remote health monitoring.

2.2.3 IoT in Home Rehabilitation

IoT is being used in many applications in day-to-day life, including sensors, home appliances, computers, smartphones, and more. Some of the most widely used applications of IoT are smart homes that are equipped with sensors and actuators as well as wearable medical devices sensors such as electroencephalogram (EEG), electromyogram (EMG), electrocardiogram (ECG), oxygen saturation, and body temperature.

The sensors and actuators are then connected through a wireless network to fetch real-time information, statistics, and measurements. Data is then processed, and the signal is then sent to the actuators, and feedback from the central computing system determines how the actuators should work. Appliances such as humidifiers, air conditioners, and oxygen generators can be controlled according to the response gathered from the central computing system of the smart home.

Senior citizens may also need regular medical attention, which otherwise can prove to be of serious concern. In addition, the cost can be reduced because the patient does not have to visit or get admitted to a hospital. They are also allowed to walk and complete daily chores with a guaranteed amount of security and privacy.

2.3 CONCLUSION

This chapter provides a detailed discussion about architecture, some important protocols used between wearables and IoT devices, and IoT sensors used in healthcare devices. We discussed the current role of IoT, along with its growth and future aspects in changing data environment.

The environment and the security issues, along with safety measurements required for using IoT in healthcare, have also been discussed. Over the past decade, the use of IoT in healthcare devices has been increased in numerous ways, especially in wearable healthcare devices and telehealth mobile applications for self-health monitoring. IoT-based healthcare devices play a major role in regular health monitoring of some common diseases, including blood sugar and blood pressure, without visiting a doctor on regular basis. Some of the self-health and wellness monitoring devices also help in the early detection of diseases. IoT-based healthcare devices also play a major role in home rehabilitation, especially for small children, individuals with disabilities, and elderly persons. However, this increased use of wearable devices and healthcare applications increases privacy and security concerns of the health data stored in these devices and applications. Also, the exchange and sharing of health data between healthcare devices, nursing centers, and hospitals increases the risk of cyber-threats, thereby demanding advanced techniques to ensure end-to-end security of the data.

REFERENCES

1. Vermesan, O., Friess, P., Guillemin, P., Gusmeroli, S., Sundmaeker, H., Bassi, A., Jubert, I. S., Mazura, M., Harrison, M., Eisenhauer, M., Doody, P. Internet of Things Strategic Research Roadmap. *Internet of Things-Global Technological and Societal Trends*, 1(2011):9–52, 2011.
2. Koop, C. et al. Future Delivery of Health Care: Cybercare. *EMBM*, 27(6):29–38, 2008.
3. Rahmani, A. et al. Smart e-Health Gateway: Bringing Intelligence to IoT-Based Ubiquitous Healthcare Systems. In *CCNC'15*, 3:20–30, 2015.
4. Shen, W., et al. Smart Border Routers for eHealthCare Wireless Sensor Networks. In *WiCOM'11*, 1–4, 2011.
5. Mueller, R. et al. Demo: A Generic Platform for Sensor Network Applications. In *MASS'07*, 6:1–3, 2007.
6. Kumar A., Salau A.O., Gupta S., Paliwal, K. Recent Trends in IoT and Its Requisition with IoT Built Engineering: A Review. *Lecture Notes in Electrical Engineering*, 526, 2019. DOI: 10.1007/978-981-13-2553-3_2
7. Malasri, K. et al. *Addressing Security in Medical Sensor Networks*, 3:7–12, 2007.
8. Ameenet, M. et al. Smart e-Health Gateway: Bringing Intelligence to IoT-Based Ubiquitous Healthcare Systems. *JMS*, 36(1):93–101, 2012.

9. Hung, X. et al. An Efficient Mutual Authentication and Access Control Scheme for WSN in Healthcare. *JN*, 6(3):355–364, 2011.
10. Prajapati B., Parikh S., Patel J. Information and Communication Technology for Intelligent Systems. *ICTIS* 2017, 1:3, 2018.
11. Baali H., Djelouat H., Amira A., Bensaali F. Empowering Technology Enabled Care Using IoT and Smart Devices: A Review. *IEEE Sens J*, 18(5):1790–1809, 2017.
12. Raghupathi W., Raghupathi V. Big Data Analytics in Healthcare: Promise and Potential. *Health Inf Sci Syst*, 2:3, 2014.
13. Zhou S., Ogihara A., Nishimura S., Jin Q. Analyzing the Changes of Health Condition and Social Capital of Elderly People Using Wearable Devices. *Health Inform Sci Syst*, 2018. https://doi.org/10.1007/s13755-018-0044-2.
14. Rault T., Bouabdallah A., Challal Y., Marin F. A Survey of Energy-Efficient Context Recognition Systems Using Wearable Sensors for Healthcare Applications. *Pervasive Mob Comput*, 37:23–44, 2017.
15. Lee H. Y. The Role of Telehealth Counselling with Mobile Self-Monitoring on Blood Pressure Reduction among Overseas Koreans with High Blood Pressure in Vietnam. *J Telemed Telecare*, 25(4):241–248, 2019 May. doi: 10.1177/1357633X1 8780559. Epub 2018, Jun 22. PMID: 29933721.
16. Rani S., Ahmed S.H., Shah S.C. Smart Health: A Novel Paradigm to Control the Chickungunya Virus. *IEEE Internet Things J*, 4662:1, 2018.

3 IoT-Based Artificial Intelligence System in Object Detection

Vishal Gupta[1] and Monish Gupta[2]
[1]Bharat Electronics Limited, India
[2]University Institute of Engineering and Technology,
Kurukshetra University Kurukshetra, India

3.1 INTRODUCTION

Automatic ship detection systems are one of the most unique Computer Vision topics [1], getting much attention in recent years [2]. Sea perception can be depicted as the convincing assertion of all oceanic exercises that sway security, economy, or nature [3]. Object Detection proof and tracking are vital for testing assignments in various applications, such as surveillance. Object Detection proof incorporates discovering the edges of ships in a video frame. Object tracking is the path to finding an inquiry or diverse inquiries after some time using a camera. The capable computer's vision, the openness of high caliber and cheap camcorders, and the growing necessity for robotized video examination have made a great deal of energy for questioning tracking counts [4]. There are three steps in video assessment, the affirmation of having captivated moving articles, the following of such from every single bundling, and the assessment of item tracks to see their direction [5]. A few works have zeroed in on making a changed zone and following tallies that confine the need for management. They usually utilize a moving article work that assesses each hypothetical item arrangement with the strategy of accessible disclosures without to unequivocally deal with their information association [6,7].

In such a way, video insight all around utilizes sensors to collect data from the earth [8]. In a normal perception framework, these camcorders are mounted in settled positions or on compartment tilt contraptions and they send video streams to a specific region, called a checking room. By that point, the video streams are seen and show up and are taken after by human chiefs. Regardless, the human chiefs may confront numerous issues, while they are viewing these sensors [8,9]. One issue is an immediate consequence of the way that the chief should examine through the cameras, as the dubious request moves between the limited fields of perspective of cameras and ought not to miss some other difference while taking it. Along these lines, surveillance structures must be motorized to upgrade the execution and remove administrators' mistakes [10].

DOI: 10.1201/9781003140443-3

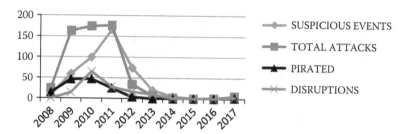

FIGURE 3.1   The statistics events from 2008–2017.

Be that as it may, there are a few weaknesses related to the utilized sensors' information [11]. In spite of the fact that the air engendering attributes for the long-wave infrared range are better than other obvious and infrared frequencies [12], when all is said and done, the air spread misfortunes boundary the scope of the electro-optical sensors to just a couple of kilometers. Further, information handling for the programmed insight era is very trying in maritime conditions. Figure 3.1 shows suspicious activities in the past few years and Table 3.1 shows a summarization of Different Techniques used by Authors

- Activity which privateer gathers unequipped for additionally privateer operation.

3.2   PROPOSED METHODOLOGY

For object recognition in maritime conditions, sensors information handling, each edge of the sensors, video flow is considered freely without considering fleeting data. The general arrangement of object identification approaches in sea. It is comprised of three primary steps, object discovery, background subtraction, and a closer view division, which are discussed below.

3.2.1   OBJECT DETECTION

There are three fundamental methodologies for recognition, (1) projection-based, (2) area-based, and (3) a Hybrid approach. Figure 3.2 indicates seven cases of sea images with a ship. Because of a variety of accessible signs and also challenges, we utilized these images as cases to exhibit the qualities and shortcomings of various recognition techniques. Figure 3.3 shows the ship in the sea.

$$xcos(\theta) + ysin(\theta) = \rho \qquad (3.1)$$

$$H(\theta, \rho) = \iint_{x,y} (1 - \delta(I(x, y)))\delta(xcos(\theta) + ysin(\theta) - \rho)dx\ dy \qquad (3.2)$$

$$R(\theta, \rho) = \iint_{x,y} I(x, y)\delta(xcos(\theta) + ysin(\theta) - \rho)dx\ dy \qquad (3.3)$$

TABLE 3.1

Summarization of Different Techniques Used by Authors

		EO Sensor		Scene		Object Detection			Object Tracking			
	Year	Infrared	Visible	On shore	Open Sea	Horizon detection	Static background	Foreground segmentation	Horizon detection	Registration	Dynamic background	Foreground tracking
Bhanu [59]	1990	YES										
Sumimoto [60], [61]	1994		YES				YES	YES			YES	
Strickland[62]	1997		YES									
Smith [63]	1999	YES					YES					
Broek [64], [65,66]	2000	YES				YES	YES	YES	YES		YES	YES
Voles [67]	2000			YES								YES
Caspi [68]	2002	YES								YES	YES	YES
Ablavsky [69]	2003										YES	YES
Mittal [70]	2004		YES								YES	
Socek [71]	2005		YES					YES				
Fefilatyev [26]	2006	YES	YES	YES	YES	YES	YES	YES				
Wang [63]	2006	YES	YES	YES			YES	YES			YES	YES
RebertInacio [9]	2007	YES	YES	YES		YES						
Schwering [60]	2007	YES	YES									
Bouma [22], [72–74]	2008	YES	YES		YES	YES	YES					
Broek [75]	2008	YES	YES				YES					
Zheng [76]	2008	YES	YES									
Blosis [1]	2009										YES	YES
Gupta[77]	2009								YES		YES	YES
Haarst [21]	2009		YES			YES	YES					
Wei[78]	2009		YES				YES	YES				
Fefilatyev [34]	2010		YES	YES	YES			YES	YES	YES		YES
Zhu [79]	2010			YES		YES					YES	
Bloisi [4]	2011		YES	YES							YES	
Hu [80]	2011		YES	YES				YES				YES
Szpak [55], [81–90]	2011		YES				YES					
Wang [19], [91–93]	2011	YES				YES	YES	YES				

(Continued)

TABLE 3.1 (Continued)

Summarization of Different Techniques Used by Authors

	Year	EO Sensor	Scene	Object Detection	Object Tracking
Broek [94]	2012	YES	YES	YES	
Ren [95]	2012	YES			
Zhang [96]	2012	YES	YES	YES	
Frost [97]	2013			YES	YES
Gershikov [18]	2013	YES	YES		
Tang [98]	2013	YES	YES	YES	YES
Blosi [99], [90]	2014	YES			
Broek [100]	2014	YES	YES	YES	
Chen [64]	2014	YES	YES	YES	YES
Tu [101]	2014	YES		YES	
Wang [61]	2014	YES			YES
Zhou [17], [102–106]	2014	YES	YES	YES	YES
Babaee [107], [108–112]	2015	YES		YES	YES
Wang [113]	2015	YES			YES
Lei, Po-Ruey [114], [115,116]	2016		YES		YES
Prasad, Dilip K., et al. [11]	2016		YES	YES	YES
Prasad, Dilip K., et al. [117], [96]	2017	YES	YES	YES	YES
Gupta, Vishal., et al. [118]	2020	YES	YES	YES	YES
Gupta, Vishal., et al. [119]	2020	YES	YES	YES	YES

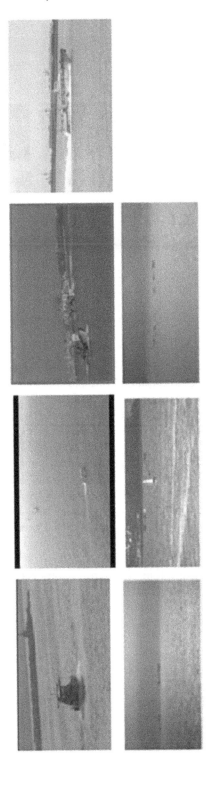

FIGURE 3.2 Here some examples of detection ship images from [55,56], [57,58].

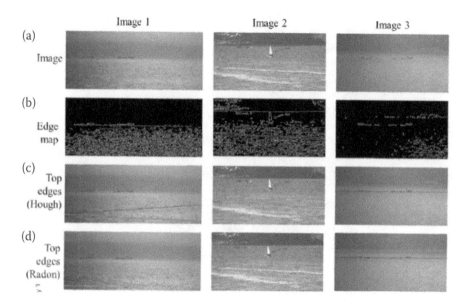

FIGURE 3.3 Ships in sea.

Like the Hough space, cells in (θ, ρ) with the most astonishing number of segments in R (θ, ρ) are the boundaries of the line. While the straightforwardness of these systems makes them famous, projective changes are delicate to preprocessing, for example, histogram leveling and disconnecting before the extraction of the edge design. Further, they can isolate the horizon in the event that it shows up as a recognizable line highlight in the edge plots. In a like manner, Figure 3.3(c, d), both Hough and Radon changes perform deficiently for picture 1 at any rate perceive the horizon in picture. Disregarding the way that the edge layout does not have critical components contrasting with the skyline, the city skyline gives a sufficient number of edge pixels equal and close to the skyline, enabling an unforgiving acknowledgment of the skyline. Strikingly, the wake makes a dull even stripe close to the goals in picture 3, which causes the discovery of the line contrasting with the wakes as well.

2) **Region-based horizon detection:** The force assortments in the area of the skyline are stand out higher from the sky or the ocean districts alone. In Figure 3.3(e,f), the areas of the skyline are depicted by generous power changes in every one of the three pictures, despite the way that the force slant itself is not enough, there is an unquestionable depiction of the skyline in picture 4. Such limited power characteristics are used for perceiving the skyline, especially in automated aeronautical vehicles [13], [14], [15]. As often as possible, the pixels in an image are assigned with a spot of sky and ocean (or ground) [16].

This is a three-phase system. The underlying advance is to use an area smoothing chairman, for instance, a formal hat channel [17], a middle channel [18], a mean channel [19], a Gaussian channel [20], or a standard deviations channel [19]. The subsequent advance is to deduce these close-by experiences with a total of Gaussian or polynomial capacities [21], [22], where each capacity is a scattering of one region, for instance, the sea regions or the sky locale.

For more troublesome depictions of the regions, for instance, a straight discriminate examination [23], surfaces [23], covariances [18], [24], and Eigen values [24], may be used. In the last step, the limit of two arranged districts is recognized as the skyline. We observe that the district-based techniques distinctively expect a priori information, for instance, sensible quantifiable depictions or AI of the example of powers at the skyline.

As opposed to the second and third steps, [22] which used a high power point to wrap up the fundamental limit of the ocean sky districts and used it as the skyline. A more enthusiastic type of this methodology used a multi-scale approach [25], [2].

3) **Hybrid techniques:** In Fefilatyev et al. [26], neighborhood measurable components were utilized; however, unequivocal portrayals of the sea and sky areas were not utilized. At that point, utilizing an arrangement of preparing images and machine learning procedures, highlights speaking to the horizon were learned specifically. Late calculations have consolidated multi-scale separating and projection-based methodologies for giving cutting edge comes about [27], [28], [29].

3.2.2 COMPARISON OF STATIC TECHNIQUES OF OBJECT DETECTION

We have demonstrated the quantitative relationship of a few static establishment procedures for object recognition. There is an expansive corpus of works identified with the establishment derivation that started from the PC Vision bundle [30]. Sea foundation can be viewed as under two conditions, colossal oceans and near port. The preliminary of dynamic establishment is, for the most part, alleviated if a long-wave infrared sensor (e.g., a forward-looking infrared (FLIR) is utilized, considering the way that it maps the temperature of the water, which is modestly uniform paying little notice to the dynamicity of water.

This covering of dynamicity jumps out at a smaller degree in close infrared and mid-wave infrared frequencies [9]. Close to thought was utilized in [31], [32], where, rather than the releasing up cutoff utilized as a bit of [16], assurance maps [31] and chi-squared extent of similarity [32] were utilized to partition the image into establishment and frontal zone zones.

Gal [33] utilized a co-event network to deal with overseeing the learning of sea and sky plans, which were deducted from the essential picture. Fefilatyev [34] utilized a Gaussian low pass sifting in a restricted strip under the horizon, trailed by a hiding slant channel to make sure that districts of high covering arrangements at long last related a breaking point figured utilizing Otsu's system [35] to get the establishment.

Multi-scale moves close, got along with low-pass channel disposing of low spatial frequencies [36], [36], [37–53], which are portrayals of establishment, have besides been discovered significant. For instance, a beat cap convolutional channel, which is a low-pass channel, utilized as a bit of a multi-stage approach appeared, apparently, to be reasonable in wake disguise [17]. Multi-scale spatio-silly advancement was utilized as a bit of [54].

Wang and Zhang [19] utilized a staggered channel and recursive OTSU approach [35] to perceive and section almost nothing and either dull or stunning focuses from pictures with complex establishment. Despite the way that wakes and froth give off an impression of being explicit in the conspicuous arrive at pictures, they are by no means shrouded even in infrared pictures.

3.3 OBJECT TRACKING

In a significant part of the writing identified with object tracking in sea conditions, the issue of object tracking is lessened to the issue of object identification in each edge. We separate between object discovery calculations and object tracking calculations in that the latter utilizes (i) temporal data across edges, e.g. optical flow, and (ii) utilizes dynamic background subtraction calculations for a robust modeling of the background. An ordinary pipeline for maritime object tracking appears in Figure 3.4. Below, we talk about each of the modules in the pipeline

All the techniques given for all practically useful precision and recall.

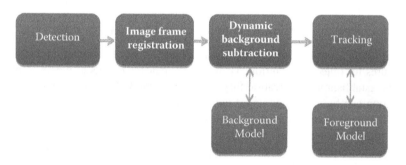

FIGURE 3.4   Object tracking steps in maritime.

TABLE 3.2

Testing results in terms of accuracy

Sr. No	Types	Total	(Percentage Correctly Classified)							
1	Butterfly	67	23.76	34.21	30.26	56.58	40.79	40.79	40.79	39.47
2	Garfield	26	56.34	36.84	36.84	0.00	26.32	26.32	26.32	26.32
3	Gramophone	35	14.23	11.11	36.11	0.00	19.44	19.44	19.44	19.44
5	Hedgehog	19	45.67	51.28	84.62	89.74	25.64	25.64	25.64	25.64
6	Ketch	78	56.12	75.76	49.49	78.79	60.61	60.61	60.61	60.61
	Total	225	123	243	215	243	215	215	215	215
	Average in %age		44.33	49.19	43.52	49.19	43.52	43.52	43.52	43.52

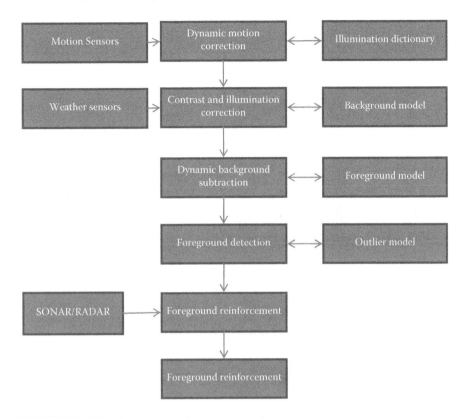

FIGURE 3.5   Flowchart of a multi-sensor processing system.

The proposed work is to take video and symbolism of the encompassing sea surface and examine it for the presence of boats, hence, conceivably empowering programmed location and following of marine vehicles as they travel in the region of the stage. The framework sends the information to the ground control through a bi-directional RF satellite connection and has its main goal boundaries reinvented during the arrangement. The depicted unit is easy, simple to send and recuperate, and does not uncover itself to likely targets. This paper also talks about the framework equipment, engineering, and calculations for visual boat discovery and tracking.

While PC Vision strategies have progressed video preparing and insight age for a few testing dynamic situations, Foreground articles, drifting, or exploring in water, including ocean vessels, little close to home boats and kayaks, floats, trash, and so forth Air vehicles, winged creatures, and fixed structures, for example, in ports, water dot, and so on qualify as a foundation. The paper investigates the chance of calculating in earlier information on a boat's shape into level set division to improve results, an idea that is unaddressed in oceanic observation issue. It is demonstrated that the created video global positioning framework beats level set-based frameworks that do not use earlier shape information, functioning admirably even when these frameworks come up short.

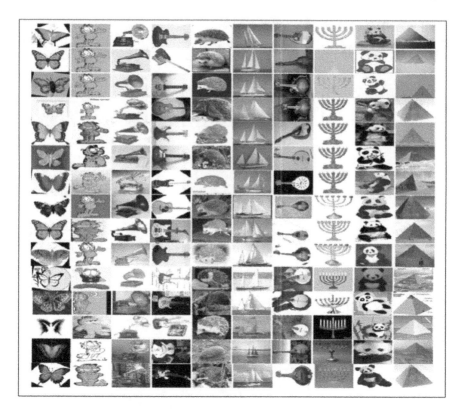

FIGURE 3.6 Images of objects used in training.

3.4 EXPERIMENTAL ANALYSIS

Here, we have shown the different images of multiple objects, and the mathematical results that we obtained. The results are considered in terms of accuracy. The intelligence IoT system is capable of distinguishing different objects. Table 3.2 shows the results in terms of accuracy.

$$\text{Recognition Rate} = \frac{\text{Number of true positives}}{\text{Number of images containing the object class}} * 100$$

3.5 CONCLUSION

Maritime video preparing issue postures truant serious preparing applications. It needs extraordinary attention to handle such difficulties. It likewise needs calculations with better flexibility to the different conditions experienced in sea situations. Thus, the field is rich with possible results of advancement in oceanic video, in terms of development. The exploration development is upheld with subjective assessment of the execution of a couple of delegate oceanic techniques Marine Dataset. The assessment shows that PC Vision frameworks can assist at sea with a reasonable turn of events. Appropriately, foundation deduction is an essential bit of ocean sensors' data planning. The foundation may be performed on one picture at some random second anticipating static foundation or may join transitory information by showing the foundation as incredible. Different object tracking methodologies used as a piece of ocean tracking have been discussed. The recognizable proof of predictabilities has ramifications for the sort of oceanic robbery and for computing the areas of assaults (Figures 3.5 and 3.6).

REFERENCES

1. Bloisi, Domenico, and Luca Iocchi. "Argos—A video surveillance system for boat traffic monitoring in Venice." *International Journal of Pattern Recognition and Artificial Intelligence*, 23.07 (2009): 1477–1502.
2. Fefilatyev, Sergiy, et al. "Detection and tracking of ships in open sea with rapidly moving buoy-mounted camera system." *Ocean Engineering*, 54 (2012): 1–12.
3. Kazemi, Samira, et al. "Open data for anomaly detection in maritime surveillance." *Expert Systems with Applications*, 40.14 (2013): 5719–5729.
4. Bloisi, Domenico, et al. "Automatic maritime surveillance with visual target detection." Proc. of the International Defense and Homeland Security Simulation Workshop (DHSS). 2011.
5. Schweitzer, Haim, Rui Deng, and Robert Finis Anderson. "A dual-bound algorithm for very fast and exact template matching." *IEEE Transactions on Pattern Analysis and Machine Intelligence*, 33.3 (2011): 459–470.
6. Blaschke, Thomas. "Object based image analysis for remote sensing." *ISPRS Journal of Photogrammetry and Remote Sensing*, 65.1 (2010): 2–16.
7. Porathe, Thomas, Johannes Prison, and Yemao Man. "Situation awareness in remote control centres for unmanned ships." Proceedings of Human Factors in Ship Design & Operation, 26–27 February 2014, London, UK.
8. Fiorini, Michele, and Jia-Chin Lin. *Clean Mobility and Intelligent Transport Systems*. Vol. 1. IET, 2015.
9. Robert-Inacio, F., A. Raybaud, and E. Clement. "Multispectral target detection and tracking for seaport video surveillance." Proc Image Vision Comput New Zealand. Hamilton, New Zealand. December (2007): 5–7.
10. Hampapur, Arun, et al. "Smart video surveillance: exploring the concept of multiscale spatiotemporal tracking." *IEEE Signal Processing Magazine*, 22.2 (2005): 38–51.
11. Prasad, Dilip K., et al. "Challenges in video based object detection in maritime scenario using computer vision." arXiv preprint arXiv: 1608.01079 (2016).
12. Chen, Chuan Cheng. *Attenuation of electromagnetic radiation by haze, fog, clouds, and rain*. Vol. 1694. No. PR. RAND Corp, Santa Monica, CA, 1975.
13. Demonceaux, Cédric, Pascal Vasseur, and Claude Pégard. "Omnidirectional vision on UAV for attitude computation." *Robotics and Automation, 2006*. ICRA 2006. Proceedings 2006 IEEE International Conference on. IEEE, 2006.
14. Ettinger, Scott M., et al. "Towards flight autonomy: vision-based horizon detection for micro air vehicles." Florida Conference on Recent Advances in Robotics. Vol. 2002, 2002.
15. Thomanek, Frank, Ernst Dieter Dickmanns, and Dirk Dickmanns. "Multiple object recognition and scene interpretation for autonomous road vehicle guidance." Intelligent Vehicles' 94 Symposium, Proceedings of the. IEEE, 1994.

16. Bhanu, Bir, and Richard D. Holben. "Model-based segmentation of FLIR images." *IEEE Transactions on Aerospace and Electronic Systems*, 26.1 (1990): 2–11.

17. Zhou, Jianjun, Haoyin Lv, and Fugen Zhou. "Infrared small target enhancement by using sequential top-hat filters." *International Symposium on Optoelectronic Technology and Application 2014*. International Society for Optics and Photonics, 2014.

18. Gershikov, Evgeny, Tzvika Libe, and Samuel Kosolapov. "Horizon line detection in marine images: Which method to choose?" *International Journal on Advances in Intelligent Systems*, 6.1 (2013): 123–130.

19. Wang, Xiaoping, and Tianxu Zhang. "Clutter-adaptive infrared small target detection in infrared maritime scenarios." *Optical Engineering*, 50.6 (2011): 067001–067001.

20. Sheng, Yunlong, et al. "Real-world multisensor image alignment using edge focusing and Hausdorff distances." Proc. SPIE. Vol. 3719. 1999.

21. vanValkenburg-van Haarst, Tanja YC, and Krispijn A. Scholte. "Polynomial background estimation using visible light video streams for robust automatic detection in a maritime environment." *SPIE Europe Security+ Defence*. International Society for Optics and Photonics, 2009.

22. Bouma, Henri, et al. "Automatic detection of small surface targets with electro-optical sensors in a harbor environment." *SPIE Europe Security and Defence*. International Society for Optics and Photonics, 2008.

23. Todorovic, Sinisa, and Michael C. Nechyba. "A vision system for intelligent mission profiles of micro air vehicles." *IEEE Transactions on Vehicular Technology* 53.6 (2004): 1713–1725.

24. Ettinger, Scott M., et al. "Vision-guided flight stability and control for micro air vehicles." *Advanced Robotics*, 17.7 (2003): 617–640.

25. Romeny, Bart M. Haar. *Front-end vision and multi-scale image analysis: multi-scale computer vision theory and applications, written in mathematica.* Vol. 27. Springer Science & Business Media, 2008.

26. Fefilatyev, Sergiy, et al. "Horizon detection using machine learning techniques." *Machine Learning and Applications, 2006.* ICMLA'06. 5th International Conference on. IEEE, 2006.

27. Prasad, Dilip K., et al. "MSCM-LiFe: multi-scale cross modal linear feature for horizon detection in maritime images." *Region 10 Conference (TENCON), 2016 IEEE.* IEEE, 2016.

28. Prasad, Dilip K., et al. "MuSCoWERT: multi-scale consistence of weighted edge Radon transform for horizon detection in maritime images." *JOSA A*, 33.12 (2016): 2491–2500.

29. Fefilatyev, Sergiy. "Algorithms for visual maritime surveillance with rapidly moving camera." 22.1 (2012): 65–72.

30. Bouwmans, Thierry. "Traditional and recent approaches in background modeling for foreground detection: an overview." *Computer Science Review*, 11 (2014): 31–66.

31. Jabri, Sumer, et al. "Detection and detection of people in video images use adaptive fusion of color and edge information." Pattern Recognition, 2000. Proceedings. 15th International Conference on. Vol. 4. IEEE, 2000.

32. Mason, Michael, and Zoran Duric. "Using histograms to detect and track objects in color video." Applied Imagery Pattern Recognition Workshop, AIPR 2001 30th. IEEE, 2001.

33. Gal, Oren. "Automatic obstacle detection for USV's navigation using vision sensors." *Robotic sailing.* Springer Berlin Heidelberg, 2011. 127–140.

34. Fefilatyev, Sergiy, Dmitry Goldgof, and Chad Lembke. "Tracking ships from fast moving camera through image registration." Pattern Recognition (ICPR), 2010 20th International Conference on. IEEE, 2010.

35. Otsu, Nobuyuki. "A threshold selection method from gray-level histograms." *IEEE Transactions on Systems, Man, and Cybernetics*, 9.1 (1979): 62–66.
36. Hou, Xiaodi, and Liqing Zhang. "Saliency detection: A spectral residual approach." *Computer Vision and Pattern Recognition, 2007*. CVPR'07. IEEE Conference on. IEEE, 2007.
37. Guo, Chenlei, Qi Ma, and Liming Zhang. "Spatio-temporal saliency detection using phase spectrum of quaternion fourier transform." *Computer Vision and Pattern Recognition, 2008*. cvpr 2008. IEEE conference on. IEEE, 2008.
38. Stauffer, Chris, and W. E. L. Grimson. "Adaptive background mixture models for real-time tracking." *Computer Vision and Pattern Recognition, 1999*. IEEE Computer Society Conference on. Vol. 2. IEEE, 1999.
39. Li, Liyuan, et al. "Foreground object detection from videos containing complex background." Proceedings of the eleventh ACM international conference on Multimedia. ACM, 2003.
40. Gupta, Kalyan Moy, et al. "Adaptive maritime video surveillance." *SPIE Defense, Security, and Sensing*. International Society for Optics and Photonics, 2009.
41. Zhong, Jing. "Segmenting foreground objects from a dynamic textured background via a robust kalman filter." Computer Vision, 2003. Proceedings. Ninth IEEE International Conference on. IEEE, 2003.
42. Koller, Dieter, Joseph Weber, and Jitendra Malik. "Robust multiple car tracking with occlusion reasoning." European Conference on Computer Vision. Springer Berlin Heidelberg, 1994.
43. L. Li, W. Huang, I. Y. Gu, and Q. Tian, "Foreground object detection from videos containing complex background," in ACM International Conference on Multimedia (2003), pp. 2–10.
44. A. Sobral, "BGSLibrary: An opencv c++ background subtraction library," in IX Workshop de VisoComputacional (WVC'2013), 2013, pp. 1–16. [Online]. Available: https://github.com/andrewssobral/bgslibrary
45. N. J. McFarlane and C. P. Schofield, "Segmentation and tracking of piglets in images." *Machine Vision and Applications*, 8.3 (1995): 187–193.
46. Z. Zivkovic and F. van der Heijden, "Efficient adaptive density estimation per image pixel for the task of background subtraction," *Pattern Recognition Letters*, 27.7 (2006): 773–780.
47. A. Elgammal, D. Harwood, and L. Davis, "Non-parametric model for background subtraction," in European Conference on Computer Vision. Springer, 2000, pp. 751–767.
48. B. D. Lucas, T. Kanade et al., "An iterative image registration technique with an application to stereo vision." *IJCAI* 81.1 (1981): 674–679.
49. P. Westall, J. J. Ford, P. O'Shea, and S. Hrabar, "Evaluation of maritime vision techniques for aerial search of humans in maritime environments," in International Conference on Digital Image Computing: Techniques and Applications, 2008, pp. 176–183.
50. S. P. van den Broek, P. B. Schwering, K. D. Liem, and R. Schleijpen, "Persistent maritime surveillance uses multi-sensor feature association and classification," in *SPIE Defense, Security, and Sensing*, 2012, vol. 3, pp. 83 920O:1–11.
51. D. Angelova and L. Mihaylova, "Extended object tracking using Monte Carlo methods," *IEEE Transactions on Signal Processing*, 56.2 (2008): 825–832.
52. N. Sang and T. Zhang, "Segmentation of FLIR images by target enhancement and image model," in *International Symposium on Multispectral Image Processing*, 3545 (1998): 274–277.
53. P. Voles, M. Teal, and J. Sanderson, "Target identification in a complex maritime scene," in *IEE Colloquium on Motion Analysis and Tracking*, 4 (1999): 99–105.

54. Yao, Zhijun. "Small target detection under the sea using multi-scale spectral residual and maximum symmetric surround." *Fuzzy Systems and Knowledge Discovery (FSKD)*, 2013 10th International Conference on. IEEE, 2013.

55. Szpak, Zygmunt L., and Jules R. Tapamo. "Maritime surveillance: Tracking ships inside a dynamic background using a fast level-set." *Expert systems with applications* 38.6 (2011): 6669–6680.

56. Saghafi, SASJ Mohammad Hassan, and Seyed Majid Noorhosseini. "Robust ship detection and tracking using modified ViBe and backwash cancellation algorithm." *Lecture Notes of the Institute for Computer Sciences, Social Informatics and Telecommunications Engineering* 117 (2014).

57. Roy, Jean. "Anomaly detection in the maritime domain." *Optics and Photonics in Global Homeland Security IV*. Proceedings of the SPIE 6945 (2008): 69450W–69450W.

58. Laxhammar, Rikard. *Anomaly detection in trajectory data for surveillance applications*. Diss. ÖrebroUniversity, 2011.

59. B. Bhanu and R. D. Holben, "Model based segmentation of FLIR images," *IEEE Transactions on Aerospace and Electronic Systems*, 26.1 (1990): 2–11.

60. Sumimoto, Tetsuhiro, et al. "Machine vision for detection of the rescue target in the marine casualty." *Industrial Electronics, Control and Instrumentation, 1994. IECON'94.*, 20th International Conference on. Vol. 2. IEEE, 1994.

61. Wang, Yu, et al. "Aquatic debris monitoring using smartphone-based robotic sensors." Proceedings of the 13th international symposium on Information processing in sensor networks. IEEE Press, 2014.

62. Strickland, Robin N., and He Il Hahn. "Wavelet transforms methods for object detection and recovery." *IEEE Transactions on Image Processing*, 6.5 (1997): 724–735.

63. Wang, ZhiCheng, et al. "Small infrared target fusion detection based on support vector machines in the wavelet domain." *Optical Engineering*, 45.7 (2006): 076401–076401.

64. Chen, Hao, et al. "Real-time automatic small infrared target detection using local spectral filtering in the frequency." *SPIE/COS Photonics Asia*. International Society for Optics and Photonics, 2014.

65. Schwering, Piet BW, Sebastiaan P. van den Broek, and Miranda van Iersel. "EO system concepts in the littoral." *Defense and Security Symposium*. International Society for Optics and Photonics, 2007.

66. Smith, A. A., and M. K. Teal. "Identification and tracking of maritime objects in near-infrared image sequences for collision avoidance." (1999): pp. 250–254.

67. P. Voles, A. Smith, and M. K. Teal, "Nautical scene segmentation using variable size image windows and feature space clustering," in European Conference on Computer Vision. Springer, 2000, pp. 324–335.

68. Y. Caspi and M. Irani, "Spatio-temporal alignment of sequences," *IEEE Transactions on Pattern Analysis and Machine Intelligence*, 24.11 (2002): 1409–1424.

69. V. Ablavsky, "Background models for tracking objects in water," in International Conference on Image Processing, vol. 3, 2003, pp. 125–128.

70. Mittal, Anurag, and Nikos Paragios. "Motion-based background subtraction using adaptive kernel density estimation." *Computer Vision and Pattern Recognition, 2004*. Proceedings of the 2004 IEEE Computer Society Conference on. Vol. 2. IEEE, 2004.

71. Socek, Daniel, et al. "A hybrid color-based foreground object detection method for automated marine surveillance." *International Conference on Advanced Concepts for Intelligent Vision Systems*. Springer, Berlin, Heidelberg, 2005.

72. Van den Broek, Sebastiaan P., et al. "Detection and classification of infrared decoys and small targets in a sea background." *AeroSense 2000*. International Society for Optics and Photonics, 2000.

73. van den Broek, Sebastiaan P., et al. "Ship recognition for improved persistent tracking with descriptor localization and compact representations." *SPIE Security+ Defence*. International Society for Optics and Photonics, 2014.

74. van den Broek, Sebastiaan P., et al. "Recognition of ships for long-term tracking." *SPIE Defense+ Security*. International Society for Optics and Photonics, 2014.

75. van den Broek, Sebastiaan P., Henri Bouma, and Marianne AC Degache. "Discriminating small extended targets at sea from clutter and other classes of boats in infrared and visual light imagery." *SPIE Defense and Security Symposium*. International Society for Optics and Photonics, 2008.

76. Y. Zheng, K. Agyepong, and O. Kuljaca, "Multisensory data exploitation using advanced image fusion and adaptive colorization," in *Signal Processing, Sensor Fusion, and Target Recognition*, 6968 (2008): 1–12.

77. K. M. Gupta, D. W. Aha, R. Hartley, and P. G. Moore, "Adaptive maritime video surveillance," in *SPIE Defense, Security, and Sensing*, 2009, vol. 7, pp. 734 609:1–14.

78. Wei, Hai, et al. "Automated intelligent video surveillance system for ships." *SPIE Defense, Security, and Sensing*. International Society for Optics and Photonics, 2009.

79. Zhu, Changren, et al. "A novel hierarchical method of ship detection from spaceborne optical image based on shape and texture features." *IEEE Transactions on Geoscience and Remote Sensing*, 48.9 (2010): 3446–3456.

80. Hu, Wu-Chih, Ching-Yu Yang, and Deng-Yuan Huang. "Robust real-time ship detection and tracking for visual surveillance of cage aquaculture." *Journal of Visual Communication and Image Representation*, 22.6 (2011): 543–556.

81. Elfes, Alberto. "Sonar-based real-world mapping and navigation." *IEEE Journal on Robotics and Automation*, 3.3 (1987): 249–265.

82. Hansen, Roy Edgar. "Synthetic aperture sonar technology review." *Marine Technology Society Journal*, 47.5 (2013): 117–127.

83. Horne, John K. "Acoustic approaches to remote species identification: a review." *Fisheries Oceanography*, 9.4 (2000): 356–371.

84. Kumar, A., Salau, A. O., Gupta, S., Paliwal, K. (2019). "Recent trends in iot and its requisition with iot built engineering: a review." *Lecture Notes in Electrical Engineering*, Vol. 526. Springer Singapore. DOI: 10.1007/978-981-13-2553-3_2

85. Ward, K. D., C. J. Baker, and S. Watts. "Maritime surveillance radar. Part 1: Radar scattering from the ocean surface." *IEE Proceedings F (Radar and Signal Processing)*. Vol. 137. No. 2. IET Digital Library, 1990.

86. Watts, S., C. J. Baker, and K. D. Ward. "Maritime surveillance radar. Part 2: Detection performance prediction in sea clutter." *IEE Proceedings F (Radar and Signal Processing)*. Vol. 137. No. 2. IET Digital Library, 1990.

87. Vicen-Bueno, R., et al. "Ship detection by different data selection templates and multilayer perceptrons from incoherent maritime radar data." *IET Radar, Sonar & Navigation*, 5.2 (2011): 144–154.

88. Pasquariello, Guido, et al. "Automatic target recognition for naval traffic control using neural networks." *Image and Vision Computing*, 16.2 (1998): 67–73.

89. Sevgi, Levent, Anthony Ponsford, and Hing C. Chan. "An integrated maritime surveillance system based on high-frequency surface-wave radars. 1. Theoretical background and numerical simulations." *IEEE Antennas and Propagation Magazine*, 43.4 (2001): 28–43.

90. Ponsford, Anthony M., Levent Sevgi, and Hing C. Chan. "An integrated maritime surveillance system based on high-frequency surface-wave radars. 2. Operational

status and system performance." *IEEE Antennas and Propagation Magazine*, 43.5 (2001): 52–63.

91. Withagen, Paul J., et al. "Automatic classification of ships from infrared (FLIR) images." *AeroSense'99*. International Society for Optics and Photonics, 1999.

92. Makantasis, Konstantinos, Anastasios Doulamis, and Nikolaos Doulamis. "Vision-based maritime surveillance system using fused visual attention maps and online adaptable tracker." *Image Analysis for Multimedia Interactive Services (WIAMIS)*, 14th International Workshop on. IEEE, 2013.

93. Baker, Simon, et al. "A database and evaluation methodology for optical flow." *International Journal of Computer Vision*, 92.1 (2011): 1–31.

94. van den Broek, Sebastiaan P., et al. "Persistent maritime surveillance using multi-sensor feature association and classification." *SPIE Defense, Security, and Sensing*. International Society for Optics and Photonics, 2012.

95. Ren, Lei, Chaojian Shi, and Xin Ran. "Target detection of maritime search and rescue: saliency accumulation method." *Fuzzy Systems and Knowledge Discovery (FSKD)*, 2012 9th International Conference on. IEEE, 2012.

96. Zhang, Dian, et al. "A visual sensing platform for creating a smarter multi-modal marine monitoring network." Proceedings of the 1st ACM international workshop on Multimedia analysis for ecological data. ACM, 2012.

97. Frost, Duncan, and Jules-Raymond Tapamo. "Detection and tracking of moving objects in a maritime environment using level set with shape priors." *EURASIP Journal on Image and Video Processing*, 2013.1 (2013): 42.

98. Tang, Da, et al. "Research on infrared ship detection method in sea-sky background." *ISPDI 2013-Fifth International Symposium on Photoelectronic Detection and Imaging*. International Society for Optics and Photonics, 2013.

99. Bloisi, Domenico D., Andrea Pennisi, and Luca Iocchi. "Background modeling in the maritime domain." *Machine Vision and Applications*, 25.5 (2014): 1257–1269.

100. van den Broek, Sebastiaan P., et al. *Recognition of ships for long-term tracking.* Bellingham, WA: SPIE, 2014.

101. Tu, Xionggang, and Jun Chen. "Infrared image segmentation by combining fractal geometry with wavelet transformation." *Sensors & Transducers*, 182.11 (2014): 230.

102. Park, Jeonghong, Jinwhan Kim, and Nam-sun Son. "Passive target tracking of marine traffic ships using onboard monocular camera for unmanned surface vessel." *Electronics Letters*, 51.13 (2015): 987–989.

103. Wang, Han, and Zhuo Wei. "Stereovision based obstacle detection system for unmanned surface vehicle." *Robotics and Biomimetics (ROBIO)*, 2013 IEEE International Conference on. IEEE, 2013.

104. Sivaraman, Sayanan, and Mohan ManubhaiTrivedi. "Looking at vehicles on the road: a survey of vision-based vehicle detection, tracking, and behavior analysis." *IEEE Transactions on Intelligent Transportation Systems*, 14.4 (2013): 1773–1795.

105. Gonzalez, Rafael C. Digital Image Processing. Pearson Education India, 2009.

106. Bao, Gui-Qiu, Shen-Shu Xiong, and Zhao-Ying Zhou. "Vision-based horizon extraction for micro air vehicle flight control." *IEEE Transactions on Instrumentation and Measurement*, 54.3 (2005): 1067–1072.

107. Babaee, Mohammadreza, and Shahriar Negahdaripour. "3-D object modeling from 2-D occluding contour correspondences by opti-acoustic stereo imaging." *Computer Vision and Image Understanding*, 132 (2015): 56–74.

108. Vespe, Michele, Harm Greidanus, and Marlene Alvarez Alvarez. "The declining impact of piracy on maritime transport in the Indian Ocean: Statistical analysis of 5-year vessel tracking data." *Marine Policy*, 59 (2015): 9–15.

109. EU Naval Force—Somalia, Key Facts and Figures. Retrieved from: ⟨http://eunavfor.eu/key-facts-and-figures⟩ (accessed 11.05.2015).

110. Hansen, J., et al. *Information domination: Dynamically coupling METOC and INTEL for improved guidance for piracy interdiction.* Naval Research Lab, Washington, DC, 2011.

111. Posada, Monica, et al. "Maritime awareness for counter-piracy in the Gulf of Aden." *Geoscience and Remote Sensing Symposium (IGARSS), 2011 IEEE International.* IEEE, 2011.

112. Mazzarella, Fabio, et al. "Data fusion for wide-area maritime surveillance." Workshop on Moving objects at Sea. 2013.

113. Wang, Yong, et al. "Aquatic debris detection using embedded camera sensors." *Sensors*, 15.2 (2015): 3116–3137.

114. Lei, Po-Ruey. "A framework for anomaly detection in maritime trajectory behavior." *Knowledge and Information Systems*, 47.1 (2016): 189.

115. Nina Ødegaard, Atle Onar Knapskog, Christian Cochin and Jean-Christophe Louvigne, "Classification of Ships Using Real and Simulated Data in a Convolutional Neural Network", IEEE Radar Conference (RadarConf), 2016.

116. S. Musman, D. Kerr and C. Bachmann, "Automatic Recognition of ISAR Ship Images." *IEEE Transactions On Aerospace And Electronic Systems* 32.4 (October 1996).

117. Prasad, Dilip K., et al. "Video processing from electro-optical sensors for object detection and tracking in a maritime environment: a survey." *IEEE Transactions on Intelligent Transportation Systems* (2017).

118. V. Gupta and M. Gupta, " Ships Classification using Neural Network based on Radar Scattering," *International Journal of Advanced Science and Technology*, 29 (2020): 1349–1354.

119. V. Gupta and M. Gupta, "Automated object detection system in marine environment." *Mobile Radio Communications and 5G Networks, Lecture Notes in Networks and Systems* 140, https://doi.org/10.1007/978-981-15-7130-5_17

4 A Value Parity Combination-Based Scheme for Heartbeat Sounds Protection

Euschi Salah, Khaldi Amine, Kafi Redouane, and Kahlessenane Fares
Computer Science Department, Faculty of Sciences and Technology, Artificial Intelligence and Information Technology Laboratory (LINATI) University of Kasdi Merbah, 30000, Ouargla, Algeria

4.1 INTRODUCTION

The recent growth and success of the Internet, as well as the availability of relatively inexpensive digital recording and storage devices [1], has created an environment in which it is easy to obtain, reproduce, and distribute digital content without compromising on the quality. This has become a major preoccupation for the multimedia content publishing industries (sound, video, and image), as there were no technologies or techniques previously that could protect the intellectual property rights of digital media and prevent unauthorized duplication [2]. Encryption technologies can be used to prevent unauthorized access to digital content. However, encryption can help protect intellectual property rights to a certain extent only because, once digital content is decrypted, it is impossible to prevent an unauthorized user from illegally reproducing it [3]. Therefore, there is need for a more advanced technology to help both establish and to prove ownership rights, adequately monitor the possible use of the content, ensure authorized access, facilitate content authentication, and prevent illegal duplication, especially with the considerable scope and potential of illegally using specific information in the field of digital imaging with respect to many practical aspects of everyday and professional life [4], including television, the Internet, audiovisual, medical, satellite imaging, and remote monitoring. This specific need attracted the attention of various research communities and industry leaders toward creating a new information concealment form, called Digital Watermarking [5], which is currently the most convenient method for addressing privacy and copyright issues and ensuring the authenticity of multimedia products. The fundamental idea of Digital Watermarking is to ensure proper creation of metadata with accurate information about the digital

DOI: 10.1201/9781003140443-4

content to be adequately protected [6], and then to carefully hide the metadata in that content. Information stored as metadata can utilize various formats, such as a character string or binary image, or even a digital image. An effective watermarking system must guarantee the imperceptibility of the hidden data [7]. The inserted mark must not affect the visual quality of the cover. A watermarking system is also evaluated for its robustness; this criterion represents the resistance after sudden changes by attacks [8]. Capacity is also a significant measure to evaluate the effectiveness of a watermark. It represents the amount of information that can be inserted into the medium; however, the larger the band size, the greater is the degradation, and that is important [9]. To enhance the security of data exchanged in telemedicine, we propose in this work a blind watermarking scheme for heartbeat sounds protection. The watermark consists of patient information and the acquisition data. Thus, during the extraction process, a comparison between the patient information and the watermark will indicate whether the data has been altered. This will ensure that the sample corresponds to the patient indicated in the record. In this work, we propose two watermarking schemes for audio files; our goal is to find a compromise between capacity and imperceptibility to hide as much data as possible while minimizing file degradation. Our first experiment is to apply our watermarking schemes in the spatial domain; for this purpose, the data to be hidden are substituted with the sample values of the file according to the variants used. Then, the second experiment involves application of two watermarking schemes in the frequency domain, the Discrete Cosine Transform (DCT) coefficients of the file are calculated, and the substitution is performed using the values of these coefficients. Our last experiment applies our four insertion variants in the multi-resolution domain; for this purpose, a Discrete Wavelet Transform (DWT) is applied to the container and the data are inserted by substituting the least significant bit of the coefficients obtained. For each insertion domain and for each variant, the distortion measurements are calculated (between the original and watermarked files) to have a more objective idea of the distortions that have occurred in the container. This also makes it possible to determine the imperceptibility rate obtained for each variant. In addition, the concealment capacity of the two schemes varies according to the insertion domain. We could, therefore, with a summary statement (capacity/imperceptibility) determine for each insertion domain, the most suitable variant that offers the best compromise between capacity and imperceptibility. This chapter is organized as follows. In Section 2, we present the various recent data hiding works applied to audio files in the spatial, frequency, and multi-resolution domains. Section 3 is dedicated to the description and presentation of our two proposed schemes, as well as their functioning for each insertion domain. The experiments and performance evaluation of each scheme are presented in Section 4. The conclusion of this chapter and the perspectives are discussed in Section 5.

4.2 RELATED WORKS

The watermarking techniques applied to audio files are related to the different representation spaces (domains), and each insertion domain has different watermarking schemes [10]; this domain can be either a spatial domain or a transformed domain.

4.2.1 SPATIAL DOMAIN AUDIO WATERMARKING METHODS

Spatial methods allow the insertion of a watermark directly into the data samples and do not require any transformation [11]. These are simple and inexpensive methods, in terms of calculation time, as they do not require a prior processing step [12]. They are dedicated to real-time watermarking required in low-power environments.

- Hosny *et al.* [13] proposed a watermarking scheme where two successive samples are selected, and then the bits are integrated into the 3rd and 4th position of the least significant bit of the selected sample. An optimization algorithm is then used to minimize the error generated by the hiding process.
- Kundu and Kaur proposed a watermarking scheme [14] where the message to be hidden is encrypted by Vigenere Square Encryption Algorithm. Then, these data are incorporated into the deep layers by using the modified LSB method. A transposition encryption is then used to encrypt the audio file.
- Pugazhenthib and Devi present a watermarking method for embedding data in an audio file [15] using the LSB integration technique. The audio file is converted into a bit stream. The first bit of the private message to be properly integrated is carefully inserted in the least significant bit of the first selected sample. The complex process is iterated until the message bits are fully inserted into the audio file.

4.2.2 TRANSFORM DOMAIN AUDIO WATERMARKING METHODS

Specific areas of transformation have been extensively explored in the context of coding and efficient compression, and many research results can be applied to Digital Watermarking [16]. Mapping in a specific transformation domain, such as DCT or DWT, generally serves two distinguishing purposes. It decorrelates the high values of the initial sample and it concentrates the potential energy of the original signal into only a few coefficients [17].

- In the innovative approach proposed by Irawati *et al.* [18], an LWT is carefully applied and then a local sub-band is properly selected and trans-formed from the time domain to the frequency domain by the DCT. Each local frame of the complex DCT coefficients is decomposed into an ortho-gonal matrix and a triangular matrix by QR matrix decomposition. The wa-termark is then inserted into a triangular matrix by accurately quantifying the accurate data at a possible value precisely corresponding to the quantizer contained in the watermark.
- Jayarani *et al.* proposed [19] a zero watermark scheme for watermarking audio files. The complex algorithm is religiously based on the Short Time Fourier Transform (STFT). The watermark is efficiently generated using the

STFT phase spectrum, which preserves the remarkable originality of the private file. The generated watermark is then stored in an extensive database to carefully check if the desired signal is altered or not.

- Karajeh and Maqableh proposed a watermarking scheme [20] combining DCT and functional Schur decomposition. A DCT is first applied to the audio signal, and then the functional Schur decomposition is properly applied to the mid-frequency band of the complex DCT coefficients. The watermark bits substituted the LSBs of the diagonal elements of the resulting triangular matrix S. The use of DCT increases robustness and the Schur decomposition meaningfully improves the needed transparency of the integration process.

- Xue *et al.* [21] proposed a watermarking scheme combining differential singular value decomposition (SVD) and parity segmentation. A DCT is correctly applied, and then the resulting coefficients are conveniently divided into two key segments in accordance with the established order of parity. The potential energy of the matched segments will be relatively similar, so this will typically reduce necessary changes in the singular values during the hiding process. This will instantly make it possible to achieve a considerable degree of imperceptibility.

4.2.3 MULTI-RESOLUTIONAL DOMAIN AUDIO WATERMARKING METHODS

- Saadi *et al.* [22] proposed a watermarking scheme that combines DCT and DWT; a DWT is applied after framing the signal, and then a DCT is applied to each frame. Each frame is divided into two segments by sub-sampling to get a significant correlation. To adequately increase watermark security, Arnold's transformation is performed on the private message to be properly integrated.

- Kaur and Dutta proposed a watermarking algorithm [23] that focuses on data concealment using the SVD transformation of the audio signal in the DWT domain. Each bit of the watermark is hidden in the audio signal using the average quantization, to allow for the insertion of a large amount of information while maintaining a good sound quality of the signal.

- Hu *et al.* proposed [24] a novel method of audio watermarking, in which a binary integration is carefully performed in the low frequency approximation sub-band, where a three-level lifting wavelet transform (LWT) is first applied. The sorted sequence of approximation coefficient quantities is then remodeled according to the watermark's predicted bit rate.

- Jiang *et al.* proposed a novel method of audio watermarking [25] in which the watermark is efficiently generated using the chaos system and then encrypted. In this innovative approach, the local frames are then divided in an adaptive way. This established division is typically based on the global characteristics of the appropriate frame and is efficiently performed by direct synchronization invariably using a corresponding frame sequence number.

4.3 PROPOSED APPROACHES

4.3.1 SPATIAL DOMAIN

The proposed approach is a specific combination of matrix coding and extensive LSB correspondence (LSBM). Matrix coding typically allows us to ensure that the secret message is incorporated with a minimum number of changes. The active LSB correspondence mitigates the even number effect (PoV) of the LSB replacement method (LSB). The proposed data hiding schemes are based on the difference in sample values. The sample values are carefully compared to each other according to the specific variant properly used. The watermark to be carefully inserted is a specific sequence of 0 and 1; these necessary data substitutes the LSB of the sample values typically depending on the specific variant used accordingly. In the first proposed watermarking scheme, three sample values are properly used to carefully hide two possible bits. Let R1, R2, and R3 be three values corresponding to the three successive sample values and X, Y two bits of the secret message. For the second variant, the values A, B, and C correspond to the last significant bits of three successive sample values R1, R2, and R3. In the second scheme, two sample values are used to hide a bit. The difference between the two sample values is calculated. Let R1 and R2 be two values corresponding to two successive sample values and X the bit to be integrated. For the second variant, the values A and B correspond to the last significant bits of the successive sample values P1 and P2.

4.3.2 FREQUENCY DOMAIN

In the proposed approach for the frequency domain, a DCT is carefully applied to the audio frames. After the thresholding and quantification step the watermark is carefully inserted into the complex DCT coefficients to get the watermarked file.

4.3.2.1 The Discreet Cosine Transform (DCT)

DCT is an audio transformation technique [26] that transforms data from the spatial domain into a transformation domain. This linear transformation transfers a matrix of n elements into another matrix of n coefficients as a cumulative sum of cosine functions of different frequencies. The matrix is, thus, divided into three frequency bands: low frequency (LF), medium frequency (MF), and high frequency (HF). The largest quantity of energy is concentrated in the LF band, while the smallest quantity of energy is concentrated in the HF band. Embedding the watermark in the HF band may affect the audibility of the signal, while embedding the watermark in the HF band may affect the quality of the watermarked bits. Therefore, in our approach, the watermark is integrated into the MF band because it is the one with least sensitive to noise. After calculating the DCT by formula 1, a thresholding is performed to increase the compression efficiency. If the absolute values of the (non-zero) coefficients of the DCT matrix obtained are below a certain threshold, these coefficients will be eliminated (set to zero).

$$F_{u,v} = \frac{1}{\sqrt{2N}} \, C_u \sum_{x=1}^{N} f(x). \, \cos\left[\frac{(2x+1)u\pi}{2N}\right] \tag{4.1}$$

Where $C_u = \begin{cases} \frac{1}{\sqrt{2N}} \text{ if } u = 0 \\ 1 \text{ if } u > 0 \end{cases}$

4.3.2.2   Quantification Process

Quantification of the obtained DCT coefficients is performed. The quantification of each block groups together the sets of similar values. This minimizes the data required to represent the audio file. A single quantum value, thus, represents a range of values [27]. To calculate the quantization matrix, the value of the quantization Q, (Q represents the number of bits required to code each element of the DCT matrix) is selected. The two MAX and MIN values of the DCT matrix (DCTmax, DCTmin) are determined. Then, the DCTQ matrix is calculated by the following formula:

$$DCTQ(i, j) = \frac{(-1 + 2^Q)(DCT(i, j) - DCTmin)}{DCTmax - DCTmin} \tag{4.2}$$

4.3.2.3   Integration Process

In the first complex scheme, the parity of three successive coefficients is exactly calculated and then compared to the possible bits to be carefully hidden. In the specific case of used inequality, the parity of a determined constant is then modified to sufficiently satisfy equality according to the specific rules in Table 4.1. For the first variant, R1, R2, and R3 typically represent three successive coefficients and X, Y two possible bits of the private message to be efficiently hidden. For the second developed variant, the values A, B, and C typically represent the possible remains of the entire division on two of the possible values R1, R2, and R3. In the second proposed scheme, the parity of two successive coefficients is accurately calculated;

TABLE 4.1
First Proposed Insertion Scheme for the Spatial Domain

Variant 1		Variant 2	
Condition	Action	Condition	Action
X= (R1 – R2) % 2 Y= (R1 – R3) % 2	No change required	X= (A $\oplus$ B) % 2 Y= (A $\oplus$ C) % 2	No change required
X≠ (R1 – R2) % 2 Y= (R1 – R3) % 2	Change R2 component	X≠ (A $\oplus$ B) % 2 Y= (A $\oplus$ C) % 2	Change R2 component
X= (R1 – R2) % 2 Y≠ (R1 – R3) % 2	Change R3 component	X= (A $\oplus$ B) % 2 Y≠ (A $\oplus$ C) % 2	Change R3 component
X≠ (R1 – R2) % 2 Y≠ (R1 – R3) % 2	Change R1 component	X≠ (A $\oplus$ B) % 2 Y≠ (A $\oplus$ C) % 2	Change R1 component

the possible substitution of one bit is performed by properly executing one of the established rules in Table 4.4. For the first variant, R1 and R2 accurately represent two successive coefficients and X the specific bit of the private message to be typically integrated. For the second variant, the values A and B represent the remains of the entire division on two of the values R1 and R2.

4.3.3 MULTI-RESOLUTION DOMAIN

In the proposed approach for the multi-resolution domain, a DWT is calculated, and then the bits of the secret message are integrated into the obtained coefficients. An IDWT is then applied to generate the watermarked file. Before integrating the watermark, a DWT is applied. The wavelet transformation decomposes the audio signal into approximation and detail components [28]. Approximation coefficients (AC) are low-frequency components, and detail coefficients (CD) are high-frequency components [29]. After obtaining the AC and CD components, the substitution is performed in the obtained AC coefficients. In the first elaborate scheme, as with the frequency approach, the parity of three successive coefficients is carefully calculated and then compared to the possible bits to be integrated. In the exceptional case of inequality, the parity of a complex coefficient is then modified to sufficiently satisfy the fundamental equality agreeing with the specific rules in Table 4.5. For the first variant, R1, R2, and R3 typically represent three successive constants of the necessary elements CA and X, Y two possible bits of the private message to be carefully hidden. For the second variant, the values A, B, and C represent the remains of the entire division on two of the values R1, R2, and R3. In the second proposed scheme, the parity of two successive coefficients of the necessary CA elements is accurately calculated; the possible substitution of one bit is performed by properly executing one of the specific rules in Table 4.6. For the first variant, R1 and R2 accurately represent two successive coefficients of the essential elements and X the possible bit of the private message to be carefully hidden. For the second variant, the values A and B represent the remains of the entire division on two of the values R1 and R2.

4.4 EXPERIMENTS AND RESULTS

In this section, the proposed watermarking techniques are tested on a large database of heartbeat sound files [30]. The data were gathered from two sources: 176 patient files from the general public via the iStethoscope Pro iPhone app and 176 patient files from a clinical trial in local hospitals properly using the digital stethoscope, DigiScope. We correctly applied each effective approach in the three insertion domains. In what follows, Sch1V1 and Sch1V2 refer to the first and second variant of the first proposed approach, whereas Sch2V1 and Sch2V2 refer to the first and second variant of the second proposed approach.

TABLE 4.2
Second Proposed Insertion Scheme for the Spatial Domain

Variant 1		Variant 2	
Condition	Action	Condition	Action
X= (R1 − R2) % 2	No change required	X= (A $\oplus$ B) % 2	No change required
X≠ (R1 − R2) % 2	Change R1 component	X≠ (A $\oplus$ B) % 2	Change R1 component

4.4.1 PERFORMANCE MEASUREMENT METRICS

To measure and test the performance of the proposed approaches, SNR measure is used. The SNR [31] is a standard metric that accurately measures the significant amount of distinct noise traditionally added to the audio file during the water-marking process. A great value of the SNR remains a desirable feature and means that the adequate level of considerable distortion is low; the SNR is the most commonly used assessment.

4.4.2 APPLICATION OF OUR APPROACHES IN THE SPATIAL DOMAIN

To experiment our approaches in the spatial domain, the audio files of our database were watermarked, a sequence of bits representing the patient information and the acquisition data were substituted for the least significant bits of the data sample. In our first experiment, only the last bit is modified. In the second experiment, the last two bits are modified; the modification of the second bit does not affect the parity of the values, and the equations in Tables 4.1 and 4.2 remain valid. For the third experiment, the last three bits are modified according to the equations of Tables 4.1 and 4.2.

As we can see in Figure 4.1, the distortions generated are visually undetectable. To analyze the results obtained in Table 4.3, the similarity and distortion measures were calculated by comparing the original and watermarked files. The results obtained by a simple LSB substitution were also calculated to see the difference between our approach and a simple LSB substitution. Indeed, an LSB substitution systematically modifies all the values during the integration process, which increases the distortions (68.6321dB on average). Our approach, on the other hand, only changes one value in three, which reduces the distortion of the watermarked file. As we can see from the results obtained in Table 4.3, all the proposed variants generate less distortion during the integration process than the LSB method, except for the Sch1.V2 variant; this is also explained by the number of values modified.

4.4.3 APPLICATION OF OUR APPROACHES IN THE FREQUENCY DOMAIN

To experiment our approaches in the frequency domain, a DCT was calculated for each file in our database (after block division), then a sequence of bits (representing

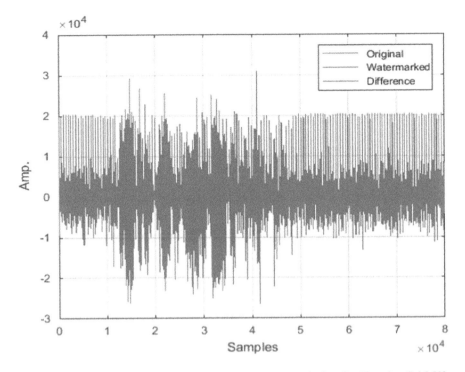

FIGURE 4.1   Difference between the original and watermarked audio file using Sch2.V2.

the patient information and the acquisition data) was substituted for the least significant bits of the quantized DCT coefficients obtained for each file, as shown in Figure 4.2. The distortion measurements are calculated between the original files and watermarked file (see Table 4.4).

As with the spatial domain, we applied a simple LSB substitution to the quantized DCT coefficients to hide the secret message. As we can see in Table 4.4, the number of values modified by a simple LSB substitution remains significantly higher than the number of values modified by our approaches. We logically obtained a higher SNR for our approaches. However, since we did not hide directly in the frame values and the quantized DCT coefficients are manipulated, the dissimulation capacity of our approaches is much less than the dissimulation capacity offered by the LSB method. Indeed, our first scheme requires three DCT coefficients for two-bit concealment, unlike the LSB method, which requires only two coefficients (for two-bit concealment).

4.4.4   APPLICATION OF OUR APPROACHES IN THE MULTI-RESOLUTION DOMAIN

To experiment with our approaches in the multi-resolution domain, each file in our database is decomposed into sub-bands (CA and DC). The CA coefficients will be used to integrate the watermark. After concealing our message (the patient

TABLE 4.3
Imperceptibility Results for the Spatial Domain Insertion

One Bit Substitution	15.000 Bits	20.000 Bits	25.000 Bits	30.000 Bits	Average
Sch2.V2	83,4430	72,8360	68,4599	63,4304	72,0423
Sch2.V1	81,7935	72,7468	69,3776	62,8172	71,6838
Sch1.V1	81,8480	71,5825	66,1580	60,0870	69,9189
LSB1	80,5659	69,9086	64,7200	59,3338	68,6321
Sch1.V2	73,9006	78,0009	65,1374	50,6372	66,9190
Two Bits Substitution	15.000 bits	20.000 bits	25.000 bits	30.000 bits	Average
Sch2.V1	77,2472	70,8889	62,9688	58,2630	67,3420
Sch2.V2	77,4098	67,4200	63,7816	58,7989	66,8526
Sch1.V1	76,8983	67,7484	62,8997	57,3613	66,2269
LSB2	74,9906	63,9234	60,5400	54,9046	63,5897
Sch1.V2	78,0023	64,3567	55,1887	53,5005	62,7621
Three Bits Substitution	15.000 bits	20.000 bits	25.000 bits	30.000 bits	Average
Sch1.V1	72,4349	64,3836	59,3592	53,9526	62,5326
Sch2.V2	75,3138	61,1884	60,9388	52,6061	62,5118
Sch2.V1	70,0062	61,0751	54,7676	49,5317	58,8452
LSB3	69,2433	60,2924	54,2560	51,2694	58,7653
Sch1.V2	69,9212	59,9488	54,9090	44,2055	57,2461

TABLE 4.4
Imperceptibility Results for the Frequency Domain Insertion

	2.000 Bits	3.000 Bits	4.000 Bits	Average
Sch2.V2	33,423895	30,1645563	21,7476577	28,4453697
Sch1.V2	32,9794132	29,990219	22,1695192	28,3797171
Sch2.V1	33,1914082	29,4845284	21,9183213	28,1980859
Sch1.V1	32,8679277	29,6838544	21,8455378	28,13244
LSB	32,390956	27,7040684	18,8971487	26,3307244

TABLE 4.5
Imperceptibility Results for the Multi-Resolution Domain Insertion

	Sch1.V2	Sch2.V2	Sch2.V1	Sch1.V1	LSB
SNR	42,962	41,918	40,89	40,089	39,718

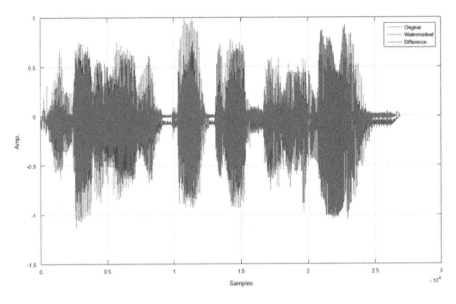

FIGURE 4.2   Difference between the original and watermarked audio file using Sch2.V2.

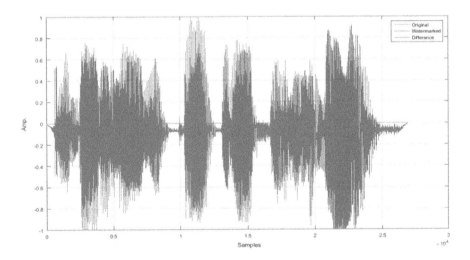

FIGURE 4.3   Difference between the original and watermarked audio file using Sch1.V2.

information and the acquisition data), the files obtained do not visually allow us to detect distortions, as seen in Figure 4.3.

To measure performance and evaluate the variants used, the measurements of similarity and distortions between the original and watermarked files were calculated (Table 4.4).

TABLE 4.6
Imperceptibility Comparison with Related Work

Spatial Domaine	LSB1	Sch1.V1	Sch1.V2	Sch2.V1	Sch2.V2	Hosny et al.	Kaur et al.	Devi et al.	
SNR	68,63	69,92	66,92	71,68	72,04	67,63	79,73	73,60	
Multi-resolution domaine	LSB1	Sch1.V1	Sch1.V2	Sch2.V1	Sch2.V2	Saadi et al.	Kaur et al.	Jiang et al.	Hu et al.
SNR	39,72	40,09	42,96	40,89	41,92	41,6001	42,419	40,60	24,63
Frequency domain	LSB1	Sch1.V1	Sch1.V2	Sch2.V1	Sch2.V2	Irawati et al.	Jayarani et al.	Karajeh et al.	Xue et al.
SNR	32,39	32,87	32,98	33,19	33,42	26,39	29,15	35,67	24,3813

To compare our approaches to a simple LSB substitution, we also implemented and tested an LSB substitution in the CA coefficients (obtained by a DWT). As with the results obtained in the spatial and frequency domains, a simple LSB substitution always generates more distortions than the proposed in Table 4.5 approaches. This is due to the number of values modified. In an LSB substitution, each coefficient has a 50% chance of being modified; this is unlike our approach, where a DWT coefficient has a 25% chance of being modified because the integration of two values is done by combining three coefficients where only one will be modified.

4.4.5 IMPERCEPTIBILITY TESTS

To accurately compare and properly set our innovative approaches with recent work, for each insertion domain, we have objectively compared the distortion results procured with those of the comparable work presented in the second section of this chapter.

As we can see in Table 4.6, the SNR obtained by our innovative approaches in the three insertion domains remains reasonable in necessary relation to the comparable work. In the first approach, the hiding process combines three values to integrate two bits, with the possibility of modifying only one value. Of these three values, only one may be modified, which typically reduces the proportional probability of possible change, unlike other methods. This typically implies less necessary modification and, therefore, less distortion of the host file; it adequately explains the good SNR rate obtained and consequently a reasonable imperceptibility. However, since our first method typically requires three possible values to carefully insert two possible bits, this positively impacts the effective approach's capacity. In the spatial domain, this can be negligible because all the data samples of the audio file are available for effective integration. However, this becomes more problematic in the frequency domain integration where the considerable number of quantized DCT coefficients is reduced.

TABLE 4.7
Robustness Comparison with Related Work

	Spatial Domaine	Frequency Domain		Multi-Resolution Domain	
	Proposed	Karajeh et al.	Proposed	Kaur et al.	Proposed
Gaussian noise	0.9728	0.9715	0.9830	0.9930	0.9978
Re-sampling	0.9218	0.9619	0.9858	0.9969	0.9963
Low-pass filter	0.9504	0.9708	0.9446	0.9429	0.98521
High-pass filter	0.9549	0.9706	0.9686	0.9068	0.9011
MP3 compression	0.9630	0.9720	0.9894	0.9348	0.94325
Cropping	0.9799	0.9558	0.9575	0.9998	0.9963

4.4.6   ROBUSTNESS TEST

To carefully evaluate the robustness of the proposed schemes (for each transform), several possible attacks were correctly executed on the watermarked samples, and the obtained results after the successful extraction of the watermark are carefully compared to the comparable works. The normalized correlation (NC) is properly applied to objectively evaluate the robustness of the watermarking scheme by properly comparing the original and the extracted watermark (after typically attacking the watermarked samples). Normally, an NC > 0.85 implies a significant similarity between the original and extracted watermark.

As we can undoubtedly see in Table 4.7, the proposed schemas typically generate reasonably robust watermarked samples against various attacks with a high-quality watermark. Habitually, in the spatial domain, if the LSBs are modified, the successful extraction of the watermark becomes impossible. However, in the proposed schemes, we can instantly recover the embedded watermark even after the LSB bit distortion. This is adequately explained by the little modification efficiently performed in the proposed approaches, which typically reduces the proportional probability of considerable change. We can, therefore, conclude from the obtained results that our innovative approaches are resistant to several malicious watermark deletion attacks while guaranteeing a reasonable SNR value.

4.5   CONCLUSION

In this chaper, two new approaches have been proposed for the watermarking of heartbeat sounds; these two approaches have been applied in three insertion domains (spatial, frequency, and multi-resolution). In the spatial domain, a specific combination between the data value samples allowed us to carefully hide two special bits by modifying only one value. The remarkable fact of modifying as few values as possible has typically allowed us to efficiently generate fewer perceptual distortions, which typically generates for us reasonable SNR results (more than 66dB). These approaches are simple and inexpensive with respect to computation time (since they do not require a prior transformation step) and can be used for real-time watermarking required in low-power environments. This protects the samples during transfer (to preserve the medical information in the file) and avoids confusion between stored samples since the extraction of the watermark guarantees that the sample corresponds to the patient indicated in the record and, therefore, avoids false diagnoses. Our approaches were then applied in the frequency domain, after the DCT calculation and quantization; the substitution of the bits to be integrated was performed by modifying the LSBs of the quantized coefficients by comparing the parity of the successive coefficients. The practical use of transforms typically makes the private message more robust to efficient compression, since it properly utilizes the similar space exploited for coding. In valuable addition, these elaborate schemes can typically provide better robustness against filtering attacks, as the marked coefficients are equitably distributed throughout the audio file. The satisfactory results observed are objectively compared to some recent work with an SNR greater than 33dB on average. However, using small audio files (which reduces the number of complex coefficients) and since our innovative approach typically requires three complex coefficients to carefully insert two possible bits, the capacity of our effective approaches is extremely limited in the frequency domain (between 4,328 and 5,770 bits per patient file). Finally, the proposed methods were properly implemented to the multi-resolution domain by substituting the complex coefficients of the local CA sub-bands properly obtained. As in the two previous domains, the parity of three successive coefficients is typically combined to properly integrate two possible bits of the confidential message. The functional decomposition into sub-bands allowed us to refine the possible isolation of the low-frequency components, which is a less sensitive insertion space while preserving the spatial content of the patient file. Imperceptibility tests sufficiently reveal suitable and satisfactory levels of SNR as carefully compared to some recent work with an SNR greater than 40dB on average for the first effective method. Contrary to the frequency domain, the large number of complex DWT coefficients typically generated is more varying, which increases the integration space of our innovative approaches. After carefully comparing our innovative approaches to the recent works in the three insertion domains, we can reasonably conclude that our second algorithm offers a good imperceptibility for insertion in the specific frequency and spatial domains. However, for future work, considerable improvements are undoubtedly a must to meaningfully improve our flexible approach to integration in the multi-resolution domain by trying to reduce the considerable distortions generated during integration. Our insertion scheme in the spatial domain is also required to be improved to progressively increase the hiding capacity, which is an essential criterion for the watermarking method.

4.6   COMPLIANCE WITH ETHICAL STANDARDS

Conflict of interest: The authors declare that they have no known competing financial interests or personal relationships that could have appeared to influence the work reported in this paper.

Ethical approval: This article does not contain any studies with human participants or animals performed by any of the authors.

REFERENCES

1. Yong Xiang, Iynkaran Natgunanathan, Dezhong Peng, Guang Hua, Bo Liu, Spread spectrum audio watermarking using multiple orthogonal PN sequences and variable embedding strengths and polarities, *IEEE/ACM Transactions on Audio, Speech, and Language Processing*, Volumemac_mac 26, 2018, Pages 529–539.
2. R. Subhashini, K. Boopathi Bagan, Robust audio watermarking for monitoring and information embedding, The Fourth International Conference on Signal Processing, Communication and Networking, Chennai, India, 2017.
3. Yong Xiang, Guang Hua, Gleb Beliakov, John Yearwood, Patchwork-based multilayer audio WatermarkingIynkaran Natgunanathan, *IEEE/ACM Transactions on Audio, Speech, and Language Processing*, Volumemac_mac 25, 2017, Pages 2176–2187.

4. Jin-Xia Yang, Dan-Dan Niu, A novel dual watermarking algorithm for digital audio, IEEE 17th International Conference on Communication Technology, Chengdu, China, 2017.

5. Hongcai Xu, Xiaobing Kang, Yajun Chen, Yilan Wang, Rotation and scale invariant image watermarking based on polar harmonic transforms, *Optik*, Volume 183, April 2019, Pages 401–414.

6. Ratnakirti Roy, Tauheed Ahmed, Suvamoy Changder, Watermarking through image geometry change tracking, *Visual Informatics*, Volume 2, Issue 2, June 2018, Pages 125–135.

7. Min-Jae Hwang, JeeSok Lee, MiSuk Lee, Hong-Goo Kang, SVD-based adaptive QIM watermarking on stereo audio signals, *IEEE Transactions on Multimedia*, Volumemac_mac 20, 2018, Pages 45–54.

8. Wu Weina, Digital audio blind watermarking algorithm based on audio characteristic and scrambling encryption, IEEE 2nd Advanced Information Technology, Electronic and Automation Control Conference, Chongqing, China, 2017.

9. Zhenghui Liu, Yuankun Huang, Jiwu Huang, Patchwork-based audio watermarking robust against de-synchronization and recapturing attacks, *IEEE Transactions on Information Forensics and Security*, Volumemac_mac 14, 2019, Pages 1171–1180.

10. Seung-Min Mun, Seung-Hun Nam, Haneol Jang, Dongkyu Kim, Heung-Kyu Lee, Finding robust domain from attacks: A learning framework for blind watermarking, *Neurocomputing*, Volume 337, 14 April 2019, Pages 191–202.

11. V. M. Manikandan, V. Masilamani, Histogram shifting-based blind watermarking scheme for copyright protection in 5G, *Computers & Electrical Engineering*, Volume 72, November 2018, Pages 614–630.

12. Mohammadreza Sadeghi, Ramin Toosi, Mohammad Ali Akhaee, Blind gain invariant image watermarking using random projection approach, *Signal Processing*, Volume 163, October 2019, Pages 213–224.

13. Ashraf A. Hosny, Wael A. Murtada, Mohamed I. Youssef, Improving LSB audio steganography using simulated annealing for satellite telemetry, 14th International Computer Engineering Conference, Cairo, Egypt, Egypt, 2018.

14. Nisha Kundu, Amadeep Kaur, A secure approach to audio steganography, *International Journal of Engineering Trends and Technology*, Volumemac_mac 44, February 2017, Pages 123–130.

15. Ramya Devi Ra, D. Pugazhenthib, Ideal sampling rate to reduce distortion in audio steganography, International Conference on Computational Modeling and Security, Tamil Nadu, India, 2016.

16. Tzuo-Yau Fan, Her-Chang Chao, Bin-Chang Chieu, Lossless medical image watermarking method based on significant difference of cellular automata transform coefficient, *Signal Processing: Image Communication*, Volume 70, February 2019, Pages 174–183.

17. S. Haddad, G. Coatrieux, M. Cozic, D. Bouslimi, Joint watermarking and lossless JPEG-LS compression for medical image security, *IRBM*, Volume 38, Issue 4, August 2017, Pages 198–206.

18. Indrarini Dvah Irawati, Gelar Budiman, Fuad Ramdhani, QR-based watermarking in audio subband Using DCT, International Conference on Control, Electronics, Renewable Energy and Communications Bandung, Indonesia, 2018.

19. Kotaro Sonoda, Aleksander Sek, Digital watermarking method based on STFT histogram, Ninth International Conference on Intelligent Information Hiding and Multimedia Signal Processing, Beijing, China, 2013.

20. Huda Karajeh, Mahmoud Maqableh, An imperceptible, robust, and high payload capacity audio watermarking scheme based on the DCT transformation and Schur

decomposition, *Analog Integrated Circuits and Signal Processing*, Volumemac_mac 99, June 2019, Pages 571–583.

21. Yiming Xue, Kai Mu, Yuzhu Wang, Yao Chen, Ping Zhong, Juan Wen, Robust speech steganography using differential SVD, *IEEE Access*, Volumemac_mac 7, Pages 153724–153733.

22. Slami Saadi, Ahmed Merrad, Ali Benziane, Novel secured scheme for blind audio/speech norm-space watermarking by Arnold algorithm, *Signal Processing*, Volume 154, January 2019, Pages 74–86.

23. Arashdeep Kaur, Malay Kishore Dutta, A blind watermarking algorithm for audio signals in multi-resolution and singular value decomposition, International Conference on Computational Intelligence and Communication Technology, Ghaziabad, India, 2018.

24. Hwai-Tsu Hu, Jieh-Ren Chang, Shiow-Jyu Lin, Synchronous blind audio water-marking via shape configuration of sorted LWT coefficient magnitudes, *Signal Processing*, Volume 147, 2018, Pages 190–202.

25. Weizhen Jiang, Xionghua Huang, Yujuan Quan, Audio watermarking algorithm against synchronization attacks using global characteristics and adaptive frame division, *Signal Processing*, Volume 162, 2019, Pages 153–160.

26. A. O. Salau, I. Oluwafemi, K. F. Faleye, and S. Jain, Audio compression using a modified discrete cosine transform with temporal auditory masking, 5th IEEE *International Conference on Signal Processing and Communication (ICSC)*, Noida, India, 2019, Pages 135–142. DOI: 10.1109/ICSC45622.2019.8938213

27. Yuanzhi Yao, Weiming Zhang, Hui Wang, Hang Zhou, Nenghai Yu, Content-adaptive reversible visible watermarking in encrypted images, *Signal Processing*, Volume 164, November 2019, Pages 386–401.

28. Y. Gangadhar, V. S. Giridhar Akula, P. Chenna Reddy, An evolutionary program-ming approach for securing medical images using watermarking scheme in invariant discrete wavelet transformation, *Biomedical Signal Processing and Control*, Volume 43, May 2018, Pages 31–40.

29. Tanya Koohpayeh Araghi, Azizah Abd Manaf, An enhanced hybrid image water-marking scheme for security of medical and non-medical images based on DWT and 2-D SVD, *Future Generation Computer Systems*, Volume 101, December 2019, Pages 1223–1246.

30. Ed King, *Heartbeat Sounds*, Stanford University, Stanford, California, United States, https://www.kaggle.com/kinguistics/heartbeat-sounds?

31. Mohammed Mahdi Hashim, Mohd Shafry Mohd Rahim, Fadil Abass Johi, Mustafa Sabah Taha, Hassan Salman Hamad, Performance evaluation measurement of image steganography techniques with analysis of LSB based on variation image formats, *International Journal of Engineering & Technology*, Volume 7, December 2018, Pages 3505–3514.

5 Sentiment Analysis of Product Reviews Using IoT

Ankur Ratmele[1,2] and Ramesh Thakur[3]
[1]IET DAVV Indore (M.P.), India
[2]STME-NMIMS Indore (M.P.), India
[3]IIPS, DAVV, Indore (M.P.), India

5.1 INTRODUCTION

Today, technology has become an important aspect of our daily lives, where the Internet plays a very crucial role. With Web 2.0, which is gaining popularity nowadays, opinion mining or sentiment analysis is one of the emerging research areas because of the huge size of reviews available on various online platforms, such as social networks, e-commerce websites, and many such forums. Previously, when there were no such online portals, or e-commerce websites, public surveys or questionnaires about the product or its features helped consumers in selecting the product. This was a very complex and tedious task, as it required finding the people who used that particular product and were willing to share their experience or their views regarding that product. Their views and experiences are known as opinions or sentiments regarding that product. Opinions could be positive, negative, or neutral based on their experience with the particular product. Earlier, it was necessary to perform the abovementioned task, but nowadays e-commerce websites and different types of online portals have made it easy for consumers to read reviews about the product and its features. A review is a sequence of texts containing the opinions of customers based on their experience. There is no fixed format for customers to write reviews. There are large numbers of reviews available for products with ever increasing trend of online shopping i.e. of e-commerce websites. Potential customers read all the reviews regarding the product, which can help them make a conclusion regarding the product and their features, whether good or bad. But with thousands of reviews, it is very difficult for the customer to read them all manually.

So there is need for developing a technique that extracts reviews from the websites and performs sentiment analysis, which can help consumers make decisions regarding the product. If this sentiment analysis is performed using an IoT device, it also leads to energy save, thereby helping the environment.

DOI: 10.1201/9781003140443-5

5.1.1 SENTIMENT ANALYSIS

Sentiment analysis, also known as opinion mining, refers to automated methods that are used to determine and anticipate people's opinion, whether positive or negative, to support or condemn, pro or against (Sammut and Webb 2017). With the growth of social media networks, sentiment analysis has become one of the most actively studied areas for brand communication campaigns, political strategies, and even people's attitudes, emotions, and evaluations.

It is one of the most active natural language processing research fields and is also extensively studied in data mining, web mining, and text mining. Sentiment analysis, also referred to as opinion mining, is the area of research that analyzes the opinions, thoughts, assessments, evaluations, behaviors, and emotions of individuals against entities, such as goods, programs, organizations, individuals, challenges, activities, subjects, and their qualities. It constitutes a large problem area. There are various names and various tasks e.g. sentiment analysis, opinion mining, opinion retrieval, sentiment mining, subjectivity analysis, impact analysis, emotion analysis, analysis of mining, etc. All these fall under the category of sentiment analysis or opinion mining.

Typical analysis issues in sentiment analysis include:

i. Extraction of functionality and recognition of elements of textual data representing user views (Golande 2016);
ii. Sentiment classification and prediction of users' attitudes and sentiments (e.g. positive or negative) (Alhajj and Rokne 2014a; Takahashi et al. 2016);
iii. Aggregation of opinions and visualization of a simplified collection of consumer opinions predicted;
iv. Automatic analysis in microblog posts of views, feelings, and subjectivity associated with a topic (Alhajj and Rokne 2014b). These tools developed for microblog sentiment analysis can also be used to identify social media data in a real-time manner.

5.1.2 SENTIMENT ANALYSIS APPLICATION

To collect public views or opinions, surveys, opinion polling, and focus groups will no longer be necessary for a company when there is an excess of such information widely accessible. However, because of the abundance of disparate sources, identifying and recording opinion sites on the Internet and distilling the information found from them remains a challenge. Usually, each site contains a large amount of opinion text that, in lengthy blogs and forum posts, is not always quickly deciphered. It would be difficult for the ordinary human reader to locate important places and to collect and summarize their opinions. Automated systems of sentiment analysis are therefore needed.

5.1.3 DIFFERENT LEVELS OF ANALYSIS

There are various levels by which sentiments are analyzed, they are as follows:

Document level: At this step, the challenge is to classify whether a positive or negative sentiment is conveyed by an entire opinion document (Pang et al. 2002). In the case of product analysis, the process will determine if the analysis represents an overall positive or negative opinion of the product. This activity is commonly regarded as the document level sentiment classification. This level of analysis suggests that each document is pinned to a single object (e.g. a single product). However, it is not applicable to documents that measure or evaluate various persons.

Sentence level: In the sentence-level analysis, each sentence is analyzed according to the sentiment whether it is positive, negative, or neutral. Typically, neutral implies no view. This scale of interpretation is closely connected to the definition of subjectivity (Wiebe et al. 1999), which separates phrases (called impartial phrases) that convey truthful facts from phrases (called subjective phrases) that convey subjective views and opinions. We should remember, however, that subjectivity is not equal to emotion, as certain factual expressions can suggest opinions. Researchers have also evaluated clauses (Wiebe et al. 2004), but the level of the clause is still not adequate e.g. even in this bad economy, Apple, Inc. is doing quite well.

Feature or Aspect level: Document-level and sentence-level reviews are not able to discover exactly what people liked and did not like. Aspect-level degree conducts fine-grained analysis. Aspect level was earlier referred to as the level of feature (feature-based opinion mining and summarization) (Hu and Liu 2004).

Therefore, there is need for developing a technique that extracts all the reviews automatically and classifies the opinion or sentiment about the feature of a product, whether it is negative or positive. Analysis of sentiments specifically analyzes views that convey or suggest positive or negative feelings.

5.1.4 OPINION

An opinion is the attitude of the individual about an object.

> *I bought a Sony TV six months ago. (2) I simply love it. (3) The video quality is superb. (4) The sound system is good. (5) However, my wife thinks it is too heavy."*

From this review, we notice a few important points. There are a variety of opinions about the Sony TV in the above review, both positive and negative. A positive opinion on the Sony TV as a whole is expressed in sentence (2). Sentence (3) provides a favorable view of the quality of the video. A good opinion on its sound is reflected in sentence (4). A negative opinion on the weight of the Sony TV is reflected in sentence (5).

Definition (Opinion): An opinion has tuples (x,o,u,v), where x is the opinion (or sentiment) aim, o is the feeling concerning the objective, u is the holder of the opinion, and y is the moment when the opinion has been expressed.

Definition (**entity**): An object is an item, service, subject, challenge, individual, association, or event. It is a pair, e: (M, N), is defined, where M is a hierarchy of parts, sub-parts, and so on, and N is a set of attributes. There also exist their own sets of features for each element or sub-part.

Objective sentiment analysis: Objective sentiment analysis means analysis of any facts in the reviews, however it is not important to analyze the facts present in the reviews.

Subjective sentiment analysis: Subjective sentiment analysis is the analysis of the attitude of the user that has written the reviews.

Definition (aspect category and aspect expression): A particular feature of the object is a feature type of an object, while an aspect term is an individual word or phrase that appears in the text implying an aspect type.

5.1.5   IoT

Any connected device represents the thing in the IoT. Usually, things are set of sensors, actuators, or a device in which a microprocessor is embedded. Things allow synchronization unite with one another, creating the necessity for machine-to-machine connectivity (M2M). By means of wireless technology, such as Wi-Fi, etc., connectivity can be of limited or wider range using mobile networks (Kumar et al. 2019). It is important to keep the cost of IoT devices low because of the vast use of IoT devices in all aspects of daily life. In addition, IoT systems, depending on the program, should be able to perform simple tasks such as data collection, M2M communiqué, and even filtering of data to a certain extent. Thus, when designing or choosing an IoT system, it is obligatory to find a balance between cost, dealing out capacity, and energy consumption. Since IoT systems constantly gather and share a vast volume of data, IoT is very closely attached to "Big Data." An IoT infrastructure usually uses approaches for managing, saving, and processing Big Data (Salau et al. 2020). In IoT infrastructures, various networks are used, such as Kaa or Thingsboard, which has become a standard practice (Kim 2017).

Thus, IoT is all about information sharing, handing out, and behaving; more formally, data can be collected or sent in several formats, from sensor readings, feedback instructions, trigger commands, to individual blogs and market records. Consequently, the data processing region is very broad. Data mining is a big approach that can be described as the process of extracting real, previously unknown, and understandable information from digitized data to improve and maximize decision-making (Wang 2006; Bramer 2007).

The most notable applications may be divided into the following groups (Talari et al. 2017):

Smart Homes: These are the conventional home appliances like refrigerators, washing machines, or light bulbs, which can be built and are capable of interacting through the Internet with each other or with registered users, providing improved system control and maintenance, as well as minimizing energy wastage. New innovations are being introduced, such as mobile home agents, mobile door locks, etc., in addition to conventional ones.

Healthcare Assistance: Novel strategies have been developed to enhance the efficiency of care provided to a patient. Without the need for a human intervention, plasters with wearable sensors will track the status of a wound and send the information to the respective doctor.

Smart Transportation: Vehicle-embedded sensors or handheld strategy and appliances when installed in the city, can make it possible to provide optimized recommendations for routes, simple booking of parking, street lighting, telematics for public transit, and prevention of accidents.

Environmental Conditions Monitoring: Wireless sensors installed across the city build the optimal infrastructure for tracking a wide range of environmental conditions. These sensors may help build an advanced weather station, with advanced humidity detection capabilities. In addition, smart sensors can also track city-wide levels of air quality and water pollution.

5.1.6 MACHINE LEARNING

It is possible to divide machine learning (ML) algorithms into four groups based on the learning styles:

Supervised Learning: Supervised learning, using algorithms such as linear regression or random forest, solves regression problems like climate forecasting, predicting life knowledge, and people growth estimation. In supervised learning, there are two levels, the training phase and the stage of testing. There must be known labels in the data sets used in the training process. The algorithms discover the relation between input data and labels, after which they attempt to predict the output labels of the data sets (Kubat 2017).

Unsupervised Learning: This deals with issues including the diminution in dimensionality used for the representation in large data, the elicitation of attributes, or the identification of hidden frameworks. In addition, supervised learning is used to cluster topics such as feedback processes, segmentation of customers, and targeted ads. In comparison to supervised learning, no labels are available in this method. Algorithms in this group aim to classify data testing trends and bunch the data or forecast potential standards (Kubat 2017).

Semi-Supervised Learning: This is a blend of the abovementioned classifications. Both labeled and unlabeled data are used. It also deals with the changes that a portion of labeled data will offer, including unsupervised learning (Mohammed et al. 2016)

The most critical question is which ML algorithm to choose from among the various types. So the answer to the above question is that we have preferred ML algorithms that are feasible for our available input datasets and the kind of application we want to run.

5.1.7 IoT USING MACHINE LEARNING

IoT devices may also be configured in many applications to activate certain measures based either on certain predefined conditions or on some input from the collected data. However, human interaction is necessary in the interpretation of the gathered data to derive valuable information for development of smart applications. IoT systems not only need to gather data and connect with other systems, but they also need to be self-sufficient. They ought to acquire measures on the basis of context and learn from their gathered knowledge.

Accordingly, the IoT system is trained to extract reviews from the e-commerce website and classify the reviews according to the positive or negative opinion after extraction.

5.2   RELATED WORK

Various methods are employed for identification of product features. In Sun et al. (2014), the tacit characteristics were derived in the sense of the inferred characteristics according to the words of the opinion and the similarities between the characteristics of the commodity. They also constructed a matrix to illustrate the correlation between the product's words of opinion and characteristics and then use a new algorithm to remove the noises in the matrix. The context and the words of thought describe the implicit features. In Pan and Wang (2011), to avoid premature pruning of the data, a new pruning method based on the grouping of features was proposed. In the grouping procedure, the approach of determining the similarities on the basis of semantics was discarded. Instead, the words of the function were found as sequences, and the correlations between them were then calculated. Authors consider that the feature words representing the same objects and the feature words representing the different objects are similar by counting the feature words in reviews. The methodology used does not ensure that the fragments are divided into a group of semantic relationships.

Xu et al. (2010) and Huang et al. (2012), suggested a model to use the Conditional Random Fields (CRFs) paradigm to define product attributes. In Jia et al. (2010), the authors concentrated mostly on categorizing the product features commented on by the customers. An unsupervised twice-clustering product-based features method of categorization was implemented. They suggested a method for categorization of two product-based attributes of semantic correlation.

In Alengadan and Khan (2018), to define the polarity of each aspect, high-frequency aspects were identified and the corresponding opinion words were derived for the aspects. The Naivè Bayes model was used to construct an algorithm for element ranking that essentially ranks the items where reviews determine the weight of the features. In Liu et al. (2012), the Naivè Bayes model was used to construct an algorithm for element ranking that essentially ranks the items where reviews determine the weight of the features.

The authors Hu and Liu (2004) used the NLProcessor linguistic parser to evaluate each review, make the most relevant contribution in the field of opinion mining, to split a text into sentences and create the part-of-speech tag for each word. The authors Ratmele and Thakur (2015) identified various data sets and techniques that are available in the area of opinion mining and sentiment analysis.

The authors Somprasertsri and Lalitrojwong (2008) and Devasia and Sheik (2016) focused on association rule mining, which makes significant use of the a priori algorithm to identify very frequent aspects of the product. The deep recursive model is used to assess the sentiment sentences in the reviews.

While research studies pertaining to social networks and the rapid developments in IoT technology have increased considerably, less emphasis has been given to the architecture of the social IoT (Soro et al. 2017). This movement will give the community further connection to "things" and, on the other hand, with more imagination, interaction, connection, involvement with people, IoT things will be socially improved. Autonomicity in the IoT is another problem in Voutyras (2014). The trend is aimed at developing autonomous IoT devices capable of managing themselves by taking into account situational awareness, knowledge, intelligence, and social behavior. In a more independent way, things will grow and act appropriately, in a more stable and smarter way. A third movement is to link places and things to online communities to encourage more connections between people and their things, such as favorite cafes, not just themselves, but also between people (Blackstock and Lea 2011). The fourth theme in social IoT is to intend network things to interact with one another independently. Research by Wakkary et al. (2017) developed a series of ceramic bowls and cups networked together to communicate with one another autonomously using IoT, called Morse things. The analysis reported that "they focus on the essence of living with IoT stuff and explore observations into the difference between objects and people that contributed to the definition of a new form of thing that is neither human-centered technology nor non-digital items in the household." In the same way, an additional study (Okada et al. 2016) of "Social Thing" is being adopted where objects have distinctive identities and networks are created individually to cooperate with each other as if they were living beings. Social things will share data about each other autonomously and take actions accordingly. The use of "Big" social data through IoT devices is a recent and significant trend in social IoT. For example, a study explored the suggestions of individuals using IoT platforms in Yao et al. (2014). Their research suggested a coherent probabilistic system by fusing knowledge to allow more detailed suggestions through user associations (i.e. the social network of users) and stuff (i.e. stuff correlations) (Kim 2017).

5.3   METHODOLOGY

The following Figure 5.1 shows the basic method used for performing sentiment analysis using ML in IOT.

5.3.1   DATASET

In this proposed technique, we require a dataset that consists of reviews. Reviews can be extracted from any e-commerce website. We used the reviews that were extracted from one of the well-known e-commerce companies i.e Amazon. Reviews extracted from the website were stored in a .csv file format. This file consists of reviews that are unstructured because there is no fixed format to write reviews on an e-commerce website; the reviews are free format. This file is sent to the IoT device for further processing.

5.3.2 IoT DEVICE

5.3.2.1 Raspberry Pi (RPi)

An embedded system has software embedded in it as a computer hardware system. It may be an isolated system or a portion of a huge system in which a microcontroller or microprocessor is programmed to execute a particular objective (Noergaard 2013). Examples of embedded devices are conventional home appliances like automated washing machines. **RPi,** a microcomputer that can provide cloud storage and Internet access, is a good tool for building IoT platforms (Molano et al. 2015).

We used the Raspberry Pi 3 Model B as a single-board computer with wireless LAN and Bluetooth communication capabilities, as shown in Figure 5.2.

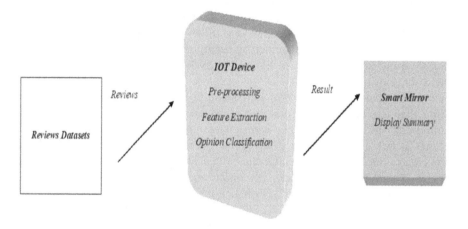

FIGURE 5.1   Method of sentiment analysis using machine learning in IoT.

FIGURE 5.2   Raspberry Pi.

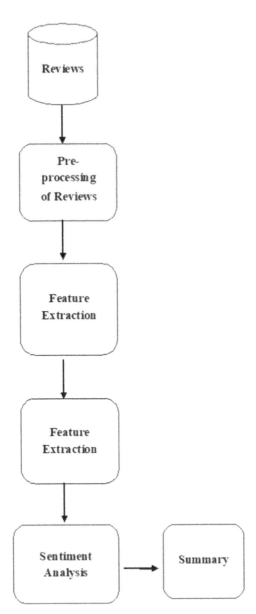

FIGURE 5.3   Architectural flow of proposed technique.

The overall architecture of the proposed techniques is shown in Figure 5.3.

5.3.3   PREPROCESSING OF REVIEWS

It is the process of cleaning and filtering of data from various stop words (words which are of no use in processing e.g. are, of, that, etc.). The most important step in preprocessing is to convert all reviews from uppercase to lower case to maintain uniformity. Reviews are then ready for further processing.

5.3.4 FEATURE EXTRACTION

This is a very important step in the proposed technique. In this Feature Extraction, parts of speech play a very important role. Stanford POS tagger is used to find Universal Dependencies and tagged reviews which are input to the proposed algorithm. Suppose sample review is "Sound of this phone is good," Universal Dependencies are nsubj(good-6, Sound-1) case(phone-4, of-2) det (phone-4, this-3) nmod(Sound-1, phone-4) cop(good-6, is-5) root(ROOT-0, good-6). This input is given to the Feat. Extraction Algorithm, as shown in Figure 5.4, and, with the help of this algorithm, the different feature is extracted

Input:
Set of Tagged Reviews (TR$_i$) ,
Set of Universal Dependencies (UD$_i$)
TR$_i$ =TR$_1$, TR$_2$, TR$_3$,......TR$_n$
UD$_{i=}$ UD$_1$, UD$_2$, UD$_3$,.....UD$_n$
Output:
Set of Feature Set (Feat)
Feat = (feat$_1$, feat$_2$,............, feat$_n$)

    1. Begin
    2.   for each TR$_i$ in TR
    3.   for each UD$_i$ in TR$_i$
    4.   if(UD$_i$ = 'nsubj')
    5.   for remaining UD$_i$ in TR$_i$
    6.   if (UD$_i$ ='conj') and arg.UD$_i$ = 'noun'
    7.   Extract feat from conj, nn
    8.   Feat.append(feat)
    9.   else if (UD$_{i=}$ 'Compound') and arg.UD$_i$ =noun && UD$_i$?'dobj'
    10.   Extract feat From compound ,nn,nnp
    11.   Feat.append (feat)
    12.   else if(UD$_i$ = 'compund' &&   UD$_{i=}$ dobj && arg.UD$_{i=}$ noun)
    13.   Extract feat from nn, compound, dobj
    14.   Feat.append (feat)
    15.   else if(UD$_i$ = nmod)
    16.   Extract feat from   nmod, nn
    17.   Feat.append(feat)
    18.   else
    19.   Extract feat from nn, amod
    20.   Feat.append (feat)
    21. Return Feat ////Integrate Feat={feat$_1$,feat$_2$....feat$_n$}

FIGURE 5.4   Feat. Extraction algorithm.

from the reviews. Our IoT machine mining is programmed to perform the above task, which results in finding the feature from reviews.

5.3.5 SENTIMENT ANALYSIS

Sentiment analysis is the process of determining the sentiment of the reviews, whether the review is positive, negative, or neutral. There are many ML techniques that are used in the determination of sentiments, such as Vader, Afinn, TextBlob, etc. But in our IoT-based ML technique, we proposed a technique that is different from others in that we have classified it according to the feature of the product. We have used TextBlob for sentiment analysis.

5.3.6 SUMMARY

Using our technique, the summary is generated, which shows the feature and its sentiment whether the feature is positive, negative, or neutral. This summary can be used by the potential customer to make a decision regarding the product. The summary is generated in the smart mirror using IoT, which also saves energy no computer system or a mobile phone is required to view the summary of the product.

5.4 EXPERIMENT AND RESULTS

In this section, the experimental works that we have accomplished to perform this research are described. Python is used to implement our algorithm. The input reviews are processed by the feature extraction algorithm, which is used to extract features from the reviews. Figure 5.5 shows a snapshot of the feature that is extracted, in which the comment (review) consists of feature camera. We have grouped this table according to feature

	Comment	Feature
0	I used to use Vivo V9.. n today got my V11.. J...	camera
1	after 4years ...using android phone.. bought o...	camera
2	Very fast delivery from flipkart. Overall phon...	camera
3	phone looks really beautiful. camera is aweso...	camera
4	best phone 10/10 great camera charging fast ...	camera
5	just one word..... awesome... from camera to b...	camera
6	Best phone ever i had used. much value for mo...	camera
7	Awesome design and display. Very good camera ...	camera

FIGURE 5.5   Snapshot of review processing.

mean reviews that consist of camera features at the top of the table, after which the remaining extracted feature with their reviews is displayed.

Features extracted by the algorithm are now classified according to their sentiments by our IoT ML technique known as TextBlob. TextBlob is ML technique that is used to identify the sentiment of the object. We also used TextBlob to classify the features according to their sentiments. Table 5.1 shows that there are five features identified by the algorithm and there are three sentiments i.e positive, negative, and neutral. Neutral also plays an important role in identification of the characteristic of the product. Figure 5.6 shows the visual representation of the feature according to the sentiment. Visual representation is shown using a two-way glass mirror. The result is displayed in the mirror so it also leads to energy saving as no separate system is needed.

TABLE 5.1
Sentiment Analysis of Reviews

Feature	Positive	Negative	Neutral
Battery	208	31	10
Camera	396	22	11
Display	230	12	7
Money	128	20	24
Performance	152	12	4

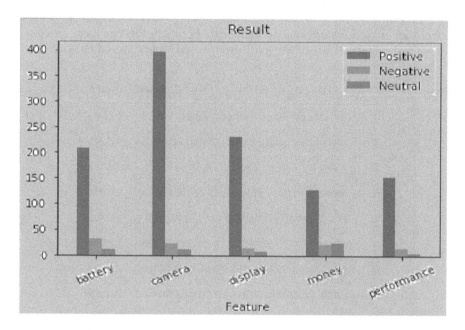

FIGURE 5.6   Visualization in glass mirror.

5.5 CONCLUSION

In this chapter, we presented how the features are extracted with the help of our algorithm, and how the sentiment analysis is done by a ML technique known as TextBlob. In this IoT-based ML technique, an IoT device is programmed, which extracts the reviews from the website and also identifies the feature from the reviews. The features are extracted with the help of our algorithm and the sentiment is identified, after which the sentiment is evaluated by using TextBlob. The summary is displayed in the smart mirror that also leads to energy saving as it does not require any computer system to read the large number of reviews. The summary helps potential customers to make a decision whether or not to buy the product. This summary is not only used by the customer but it can be also used by the manufacturer to know the sentiment of the customers for its product.

REFERENCES

Alhajj, R., and Rokne, J. (Eds.), 'Microblog sentiment analysis', in *Encyclopedia of Social Network Analysis and Mining* (Springer: New York, 2014a), pp. 893–893.

Alhajj, R., and Rokne, J. (Eds.), 'Twitter opinion mining', in *Encyclopedia of Social Network Analysis and Mining* (Springer: New York, 2014b), pp. 2259–2259.

Alengadan, B., and Khan, S. S., 'Modified aspect/feature based opinion mining for a product ranking system,' 2018 IEEE International Conference on Current Trends in Advanced Computing (ICCTAC) Bangalore, 2018, pp. 1–5.

Blackstock, M., Lea, R., and Friday, A. 'Uniting online social networks with places and things,' Proceedings of the Second International Workshop on Web of Things, San Francisco, California, 2011.

Bramer, M. *Principles of Data Mining (Vol. 180)* (Springer: London, 2007).

Devasia, N., and Sheik, R., 'Feature extracted sentiment analysis of customer product reviews,' 2016 International Conference on Emerging Technological Trends (ICETT), Kollam, 2016, pp. 1–6, doi: 10.1109/ICETT.2016.7873646.

Golande, A., 'An overview of feature based opinion mining,' in Corchado Rodriguez, J. M., et al. (Eds.), *Intelligent System*, 2016.

Huang, S., Liu, X., Peng, X., and Niu, Z., 'Fine-grained product features extraction and categorization in reviews opinion mining,' 2012 IEEE 12th International Conference on Data Mining Workshops, Brussels, 2012, pp. 680–686.

Hu, M. and Liu, B., 'Mining opinion features in customer reviews,' *AAAI*, vol. 4, no. 4, pp. 755–760, 2004.

Hu, M. and Liu, B., 'Mining and summarizing customer reviews,' Proceedings of the tenth ACM SIGKDD international conference on Knowledge discovery and data mining, ACM, 2004, pp. 168–177.

Jia, W., Zhang, S., Xia, Y., Zhang, J., and Yu, H., 'A novel product features categorize method based on twice-clustering,' 2010 International Conference on Web Information Systems and Mining, Sanya, 2010, pp. 281–284.

Kim, J. E., 'Empowering end users for social internet of things,' Proceedings of the Second International Conference on Internet-of-Things Design and Implementation, Pittsburgh, PA, 2017.

Kubat, M., *An Introduction to Machine Learning* (Springer: Cham, Switzerland, 2017).

Kumar, A., Salau, A. O., Gupta, S., and Paliwal, K., Recent trends in iot and its requisition with iot built engineering: a review,' *Lecture Notes in Electrical Engineering*, vol. 526. Springer Singapore, 2019. DOI: 10.1007/978-981-13-2553-3_2

L. Liu, Z. Lv and Wang, H., Opinion mining based on feature-level, 2012 5th International Congress on Image and Signal Processing, Chongqing, 2012, pp. 1596–1600.

Molano, J. I. R., et al., 'Internet of things: A prototype architecture using a Raspberry Pi,' in Uden, L., et al. (Eds.), *Knowledge Management in Organizations: 10th International Conference*, KMO 2015, Maribor, Slovenia, August 24–28, 2015, Proceedings' (Springer International Publishing, 2015), pp. 618–631.

Mohammed, M., Khan, M. B., Bashier, E. B. M., *Machine learning: Algorithms and Applications* (CRC Press: Boca Raton, FL, 2016).

Noergaard, T., 'Chapter 1 - A systems approach to embedded systems design,' in *'Embedded Systems Architecture* (Second Edition)' (Newnes, 2013), pp. 3–19.

Okada, M., et al., 'Autonomous cooperation of social things: Designing a system for things with unique personalities in IoT,' Proceedings of the 6th International Conference on the Internet of Things, Stuttgart, Germany, 2016.

Pang, B., Lee, L., and Vaithyanathan, S., 'Thumbs up? Sentiment classification using machine learning techniques,' Proceeding in EMNLP, 2002, pp. 79–86.

Pan, Y. and Wang, Y., 'Mining product features and opinions based on pattern matching,' Proceedings of 2011 International Conference on Computer Science and Network Technology, Harbin, 2011, pp. 1901–1905.

Ratmele, Ankur and Thakur, Ramesh, 'Statistical analysis & survey of research work in opinion mining,' *SSRN Electronic Journal, 2019*. DOI: 10.2139/ssrn.3366294.

Salau, A. O., Chettri, L., Bhutia, T. K., Lepcha, M., 'IoT based smart digital electric meter for home appliances,' *2020 International Conference on Decision Aid Sciences and Application (DASA)*, Sakheer, Bahrain, 2020, pp. 708–713. DOI: 10.1109/DASA51403.2020.9317062.

Somprasertsri, G. and Lalitrojwong, P., 'Automatic product feature extraction from online product reviews using maximum entropy with lexical and syntactic features,' in Information Reuse and Integration, 2008 IEEE International Conference on, 2008, pp. 250–255.

Soro, A., et al., 'Designing the social internet of things', Proceedings of the 2017 CHI Conference Extended Abstracts on Human Factors in Computing Systems, Denver, Colorado, 2017.

Sammut, C., and Webb, G. I. (Eds.), *Encyclopedia of Machine Learning and Data Mining* (Springer: US, 2017), pp. 1152–1152.

Sun, L., Li, S., Li, J., and Lv, J., 'A novel context-based implicit feature extracting method,' 2014 International Conference on Data Science and Advanced Analytics (DSAA), Shanghai, 2014, pp. 420–424.

Takahashi, S., et al., 'A method for opinion mining of coffee service quality and customer value by mining Twitter,' in Kunifuji, S., et al. (Eds.), *Knowledge, Information and Creativity Support Systems: Selected Papers from KICSS'2014 - 9th International Conference, held in Limassol, Cyprus, on November 6–8, 2014* (Springer International Publishing, 2016), pp. 521–528, Technologies and Applications 2016 (Springer International Publishing, 2016), pp. 633–645.

Talari, S., Shafie-Khah, M., Siano, P., Loia, V., Tommasetti, A., Catalão, J. 'A review of smart cities based on the internet of things concept,' *Energies* vol. 10, p. 421, 2017.

Voutyras, O., 'Achieving autonomicity in IoT systems via situational-aware, cognitive and social things,' Proceedings of the 18th Panhellenic Conference on Informatics, Athens, Greece, 2014.

Wang, Y., 'Integration of data mining with game theory', in Wang, K., et al. (Eds.), *Knowledge Enterprise: Intelligent Strategies in Product Design, Manufacturing, and Management* (Springer: US, 2006), pp. 275–280.

Wakkary, R., et al., 'Morse things: A design inquiry into the gap between things and Us,' Proceedings of the 2017 Conference on Designing Interactive Systems, Edinburgh, United Kingdom, 2017.

Wiebe, Janyce, Wilson, T., Bruce, R. F., Bell M., and Martin, M., 'Learning subjective language,' *Computational Linguistics*, vol. 30, no. 3, pp. 277–308, 2004.

Wiebe, Janyce, Bruce, R. F., and O'Hara, T. P., 'Development and use of a gold-standard data set for subjectivity classifications,' Proceedings of the Association for Computational Linguistics (AC L-1999), 1999.

Xu, B., Zhao, T., Zheng, D., and Wang, S., 'Product features mining based on Conditional Random Fields model,' 2010 International Conference on Machine Learning and Cybernetics, Qingdao, 2010, pp. 3353–3357.

Yao, L., et al., 'Exploring recommendations in internet of things,' Proceedings of the 37th international ACM SIGIR conference on Research & Development in Information Retrieval, Gold Coast, Queensland, Australia, 2014.

6 Saccadic Scan Path Predicting Using Convolutional Auto Encoders

Souvik Das, Shafee Anwar, and J Maiti
Department of Industrial and Systems Engineering,
IIT Kharagpur, India

6.1 INTRODUCTION

Human beings have the ability to process and understand a vast amount of visual information reaching their brain. This can be attributed to their ability in finding the most salient region of a scene and focusing on it. Eye movement consists of two main aspects, fixations and saccades. Fixation at a point brings visual information of that point, whereas saccade is the movement between the two fixations that helps in finding the most important visual location. Most of the studies on visual saliency are inspired by Center-Surround Assumption, which says that ganglion cells in the retina respond to the contrast between the center and the surrounding region. Consider, for example, the image shown in Figure 6.1.

Central and surrounding regions marked in white have the same input, hence there is no contrast between the center and the surrounding region. Therefore, neurons are not responsive to stimuli in that area. On the other hand, there is a contrast between central and surrounding regions and neurons respond to the stimuli. This property allows us to distinguish objects and find important regions in our field of view. It also reduces the computation our brain requires to examine a scene and, hence, it is widely used in computer vision models as well.

There are two different approaches in cognitive neuroscience used in computer vision when it comes to information processing scene understanding. The first one is the bottom-up approach, which integrates low-level features such as color, intensity, and orientation and builds into higher level features to predict the most salient region and is used when subjects are freely viewing the content. The second approach is the top-down, where it is task-dependent, for example, if subjects are tasked with finding the number of people in the image and the case eye-movement pattern is different than the first one. In the past two decades, extensive research has been done in this field to develop computational visual attention models, most notably by Laurent Itti, Christoff Koch, S Ulman, and others. Extensive research has been performed to predict the

DOI: 10.1201/9781003140443-6

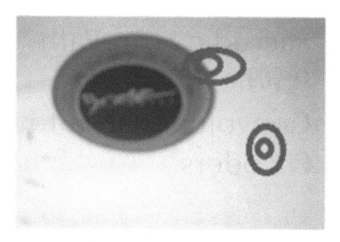

FIGURE 6.1   An example.

saliency of images but there is very little literature available in saccadic models especially in unsupervised learning. A lot of effort has been put in by the computer vision society to hand engineer the features to be used in training, such as visual attention models. Nevertheless, one always wonders whether it is possible to let the model design and extract features for itself and compare its performance with supervised learning methods and in this project, we are trying to find out the same.

Auto-encoders provide an excellent framework for semi-supervised learning as they try to approximately learn the input themselves, but since they pass through a small code layer, they are never exactly able to learn the input. This is especially useful to us as it means that an auto-encoder will learn the basic structure of the input, but high-level features, such as faces, are not learned, thus, subtracting the predicted values from actual values highlights the complicated regions of the image and this is where our fixations lie. Since the input data is in the form of images, it makes sense to use a convolutional auto-encoder, since convents perform much better than other auto-encoders on images [1]. This model incorporates a biologically inspired center-surround assumption, which finds the contrast in an image.

6.2   RELATED WORKS

Visual attention is influenced by two approaches, top-down and bottom-up. The bottom-up model is based on stimuli such as color and intensity upon which saliency is based, whereas the top-down model is influenced by high-level factors such as memory and task. Most of the work performed in this field assumes a feature integration theory proposed by Triesman et al. [2], which suggests that if there are more than one features that are needed to distinguish between two objects, then we can combine these features to identify them separately. One of the early studies in this field was performed by Itti et al. [2], based on the framework proposed by Koch and Ulman [3], where they extracted a set of features from an image, including color, intensity, and orientation and combined them to form a topographical saliency map. In that research, fixation regions were selected based on decreasing saliency in a winner-take-all fashion. Several models were built after that to improve the performance of the model implemented by Itti [2], which involved generating new features to enable the framework to extract more information from a scene. However, generating handcrafted features is always challenging, and over the past few years, unsupervised learning-based saliency models have emerged to learn feature extraction in an implicit way. For example, in [2], Xia et al. introduced sparse representation to reconstruct a central patch with all surrounding ones. The saliency of each pixel can be measured by the error between the represented and actual central patches. To model multi-layer representation and learning of the sensory cortex, more and more deep models have been involved in saliency estimation [4]. On the whole, early deep saliency models have usually exploited deep networks to learn features for distinguishing between fixated and non-fixated regions. For instance, Ku¨mmerer et al. applied the convolutional neural network (CNN) from Krizhevsky et al. to generate high-dimensional features for saliency prediction. Ehsan et al. proposed a convolutional sparse auto-encoder framework to capture the shape of learned objects extracting interpretable features to improve the performance of the saliency model. In this study, we propose a model similar to [3], with the main difference being using the convolutional auto-encoder (CAE) instead of deep auto-encoders and adding sparsity to model C-S assumption and including inhibition of return to prevent the fixation from being stuck at a single point. Employing CAE enables us to control the number of features that we want to extract, giving us control over the learning and using IOR enables us to scan the whole image and locate the salient regions.

6.3 PRE-REQUISITES

There are a few frameworks, models, and assumptions that have been used in this model, which include:

6.3.1 CONVOLUTIONAL AUTO-ENCODER

CAE is a deep learning framework used for feature extraction and is mainly on image datasets. CAE consists of **convolution filters** and **pooling layers.** Together they form a single layer. Convolution filter consists of $m \times n$ matrix and when it is convolved with an $X \times Y$ input, the resulting output is $(X - m + 1) \times (Y - n + 1)$ matrix. Consider the example

3	0	1	2	7	4
1	5	8	9	3	1
2	7	2	5	1	3
0	1	3	1	7	8
4	2	1	6	2	8
2	4	5	2	3	9

6×6

*

1	0	-1
1	0	-1
1	0	-1

3×3
filter

-5	-4	0	8
-10	-2	2	3
0	-2	-4	-7
-3	-2	-3	-16

4×4

We have a 6 x 6 input array and when we convolve it with a 3 x 3 filter we get a 4 x 4 output as shown on right. Convolution operation takes place by multiplying each element of the filter with corresponding elements on the input array starting from the upper left corner and then moving a distance known as **stride length**. In case the *stride length* is more than one, the above formula changes to: $output = ([(X - m)/l] + 1) * ([(Y - n)/l + 1)$. In case the input is three dimensional, we convolve it with a 3D filter. The number of parameters that is required for the filters to learn equals

$$total\ parameters\ (of\ 1\ layer) = filter\ size * depth\ of\ input * number\ of\ features$$
$$+ 1\ (bias)$$

In case we want to keep the dimensions of output the same as input, we can use a method known as **Padding.** In this method, first, a layer of appropriate size is added all around the input matrix and its values are set to 0. Then we convolve it with the filter keeping the dimensions intact. There are two methods to padding called *valid padding* where there is basically no padding resulting in a change in output dimension and the *same padding* output dimension is the same as the input. Convolution filters are used to extract features and are excellent edge detectors.

After the convolutional layer is a **pooling layer**. There are two methods to implement pooling operation called **max pooling** and **average pooling.** To perform pooling, we select a n x n region in the input matrix and take either the maximum value (max pooling) or average value (average pooling).

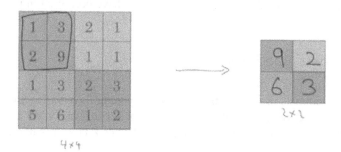

Max-pooling operation on 4 x 4 matrix gives us 2 x 2 output. Max pooling allows us to store information that might be lost if further convolutions are performed. It may store information, such as the position of the eyes.

Thus, a CAE consists of many convolution filters and pooling layers. After the final encoding layer, the output is transformed into a 1-dimensional vector that contains all the features learned from the input data. To learn an approximate representation of the input, we use this feature map and deconvolution operation is performed on it. Each deconvolution layer consists of a convolution filter and up-sampling layer. Convolution filters are same as described earlier and up-sampling

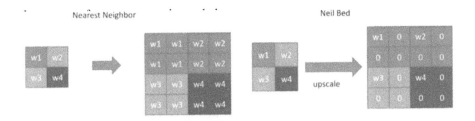

Several deconvolution layers are added and features are reduced symmetrically while the output layer size is increased till we get the same output as the input dimensions. We then train these parameters to learn the approximation of input using backpropagation method and representation errors are minimized iteratively. The following graph gives the summary of CAE:

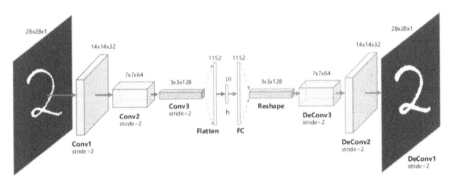

Source: Researchgate.net

6.4 PROPOSED MODEL

Using CAE framework, we propose an unsupervised learning method to predict the saccadic scan path, which adapts to the specific structure of image in question. Our model works in two steps: 1) estimation of saliency by calculating reconstruction error of trained model and 2) training the model again using samples around previous fixation to minimize reconstruction error of that region and finding next fixation. The following is the architecture of CAE:

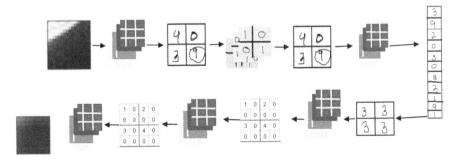

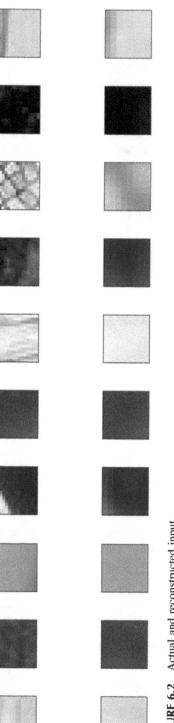

FIGURE 6.2 Actual and reconstructed input.

(b) through (f) are convolutional layers and (h) through (m) are deconvolutional layers. (a) Input surrounding of a random pixel of size $16 \times 16 \times 3$, (b) 16 convolution filters of size 3×3 with "valid padding" resulting in size $16 \times 14 \times 14$, (c) max pooling layer of size 2×2 (size becomes $16 \times 7 \times 7$), (d) convolution filter of size 2×2 with "valid padding" (size becomes $64 \times 6 \times 6$), (e) 2×2 max pooling layer (size becomes $64 \times 3 \times 3$), (f) convolving again with 128 3×3 filters with valid padding gives a 1-dimensional coding layer of size 128 (i.e. g), (h) 2×2 upsampling layer (size becomes $128 \times 2 \times 2$), (i) 64 3×3 convolution filters with "same padding" (resulting in size $64 \times 2 \times 2$), (j) upsampling layer of size 2×2 (output $64 \times 4 \times 4$), (k) 16 convolution filters of size 3×3 with "same padding" (output is $16 \times 4 \times 4$), (l) upsampling layer of size 2×2 (size becomes $16 \times 8 \times 8$), (m) 3 convolution filters of size 3×3 with same padding giving final output i.e. the predicted center. Subtracting this predicted center with real center gives us the residual error of estimation.

1. Training Auto-encoder

To predict the saccadic scan path of an image, we perform training as well as prediction on itself in an unsupervised manner. To train our auto-encoder we take the image and select approximately 1,000 random pixels. We take a surrounding region of 16×16 around all those random pixels as well as central region of 8×8. Since a major portion of an image consists of background, large samples from the background are selected, allowing for better reconstruction of the background. Due to small portion of foreground, not much sample data is collected from it. Hence, there will be a large reconstruction error owing to saliency of that region. This property is highlighted in Figure 6.3. We feed this data into auto-encoder with surrounding regions as input (a) and central regions as output (n'). We perform convolution operations on the input and keep reducing the size of each hidden layer. The system learns the filters using the back propagation method. The number of filters keeps increasing in each hidden layer. Each layer consists of a convolutional filter, which learns features, such as edges, etc., and the max pooling layer, which stores a max

Layer (type)	Output Shape	Param #
input_2 (InputLayer)	(None, 16, 16, 3)	0
conv2d_7 (Conv2D)	(None, 14, 14, 16)	448
max_pooling2d_3 (MaxPooling2	(None, 7, 7, 16)	0
conv2d_8 (Conv2D)	(None, 6, 6, 64)	4160
max_pooling2d_4 (MaxPooling2	(None, 3, 3, 64)	0
conv2d_9 (Conv2D)	(None, 1, 1, 128)	73856
up_sampling2d_4 (UpSampling2	(None, 2, 2, 128)	0
conv2d_10 (Conv2D)	(None, 2, 2, 64)	73792
up_sampling2d_5 (UpSampling2	(None, 4, 4, 64)	0
conv2d_11 (Conv2D)	(None, 4, 4, 16)	9232
up_sampling2d_6 (UpSampling2	(None, 8, 8, 16)	0
conv2d_12 (Conv2D)	(None, 8, 8, 3)	435

Total params: 161,923
Trainable params: 161,923

FIGURE 6.3 Total number of trainable parameters are 161,923.

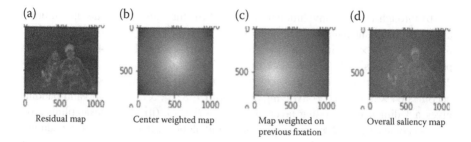

(a)   (b)   (c)   (d)

Residual map   Center weighted map   Map weighted on   Overall saliency map
                                       previous fixation

FIGURE 6.4 Various image map.

value in a 2 × 2 region. The model tries to learn the central region given the surrounding input. Following is the actual and reconstructed input:

As evident from the above figure, the auto-encoder tries to learn the basic
structure of the input feed. A summary of the auto-encoder is as shown in
Figure 6.3.

In our experiment we have taken 1,000 samples of size 16 x 16 hence total input
parameters are 256,000 and the number of parameters to be trained is ~161,000. Hence
this model can be trained. After training our encoder, we now create a saliency map.

2. Saliency Map

To create our saliency map, we find the reconstruction error by measuring the
difference between actual and predicted central patches. Reconstruction error is
measured for central patches around each pixel of the image by feeding the network
with a surrounding patch around all pixels. This gives us a residual map of the size
of input image. Results from experiments performed in [3] show that there is a
strong tendency for generating saccades of small amplitudes. To incorporate this
feature, we have generated weights centered toward previous fixation regions.
Results also show that fixations are biased toward center of images, therefore we
have also included weights at the center, thus not allowing fixations to move far
away from center. Thus overall saliency map is calculated as:

$$residual\ map = \|\ actual\ central\ patch - predicted\ central\ patch\ \|$$

$$center\ weighted\ map = 1 - distance\,(x - x_{center},\ y - y_{center})/sum\ of\ total\ distances$$

$$fixation\ distance\ map = 1 - distance\,(x - x_{old\ fix},\ y - y_{old\ fix})$$
$$/sum\ of\ total\ distances$$

$$saliency\ map = center\ weighted\ map * fixation\ distance\ map * residual\ map$$

3. **Saccadic Scan Path Generation**

We have now created a saliency map of an image. To generate a saccadic scan path, we set our first fixation as center of the image. Also, a fixation weighted map is set to the center. Now, we select 1,000 random surrounding and central regions and train our auto-encoder. Then we create a residual map by taking each pixel and creating a surrounding region around it. We then predict the central patch using our trained model. Once we have calculated the residual error between actual and predicted center of each pixel, we create a residual map by putting the error value of reconstruction of central patch at each corresponding pixel, creating residual map. Combining it along with the center-weighted map and the fixation distance map gives us an overall saliency map. Position of the maximum value of saliency map gives us the fixation point.

After finding our first fixation point, we updated the parameters of the auto-encoder by generating new samples and training on them. Half of these samples (500) were taken from the region surrounding the fixation point we found and the rest are random. This allows the auto-encoder to learn the region surrounding previous fixation so that the reconstruction error is reduced, thus reducing the probability of fixation being stuck at a single point. However, during the experiment we found that even though overall residual error reduces in each iteration, the model proposed by Chen et al. [3] was not able to learn about the region surrounding previous fixation points. Thus we use another method widely used in computer vision known as Inhibition of Return (IOR). To implement IOR, we set the residual error of region surrounding the previous fixation to 0. Implementing IOR gives us improved performance as seen below:

6.5 EXPERIMENT

The experiment was carried out using a standard open source MIT dataset [5] containing 1,003 images and eye tracking data of 15 viewers. Images included were natural as well as artificial. As our experiment requires only a single image for training as well as a prediction on its saccadic scan path, we only had to select a few images to check the performance of our model. The experiment was performed using Google virtual machine on their Google coloboratory platform [6], which provides us with 12 GB of RAM and 50 GB of disk space. We used machine learning library from Keras [7] to implement our CAE model. We added the sparsity constraint in our model while training the data so that fewer neurons fire in the hidden layer, thus giving us the ability to incorporate a C-S assumption. The activity regularizer was set to $L1 = 10e - 5$ where L1 is the mean absolute deviation. There were a total of 3 encoding and 3 decoding layers with "valid padding" in encoding layers and "same padding" in decoding layers. Size of the convolution filter was either 3×3 or 2×2 (2×2 filter was used to keep the dimensions of output whole numbers). The size of pooling and upsampling layer used was 2×2 and activations used were sigmoid

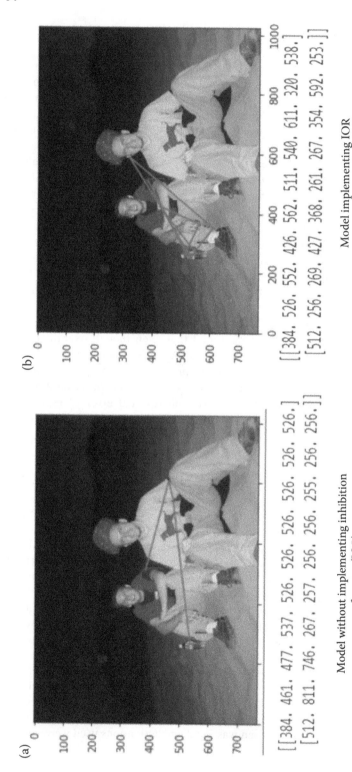

FIGURE 6.5   Implementation of IOR.

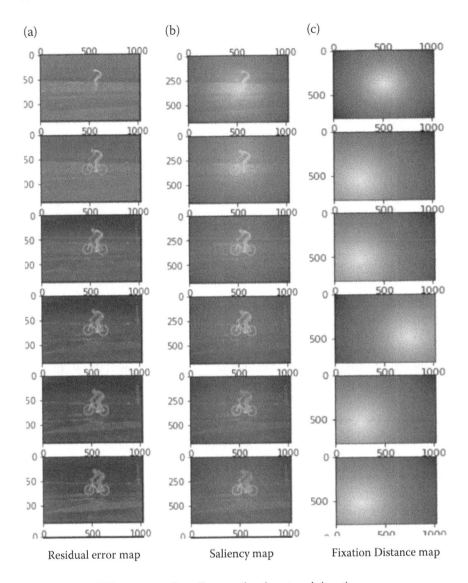

Residual error map   Saliency map   Fixation Distance map

FIGURE 6.6 Different maps for saliency estimation at each iteration.

function in the last layer and ReLU in all other layers. We tried different optimizers for training our model including "adam," "adadelta," and "stochastic gradient descent," but SGD gave better training performance over other methods. The learning rate of the optimizer varied between 0.001 and 0.1 with very high value of learning rate causing overfitting and very small values leading to slower learning. A learning rate of 0.01 was found to be ideal. Error measurement was using binary cross-entropy error; however, as pointed out in [3] this was not a very good measurement of learning, so we had to keep

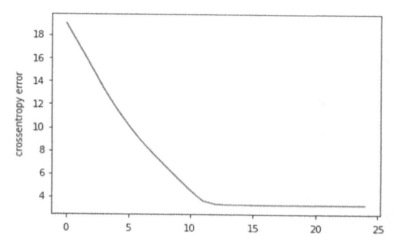

FIGURE 6.7   Cross entropy error.

checking the least square error (called L2 error) between actual and re-constructed output. The model was trained for 100 epochs and batch size of 256. After training our model, we estimated the first 10 fixation points and found that fixations were more inclined toward the edges as it had highest re-construction error implying that our model was learning correctly. Results of the experiment are discussed in the following section.

6.6   RESULTS AND CONCLUSION

Results of saliency prediction after training each time with additional training sets gave us residual and saliency maps as shown in Figure 6.6

While training the dataset using back propagation and stochastic gradient descent, the following learning curve was observed:

Cross entropy error is high initially, but decreases linearly with each epoch and becomes constant. Similar trend was observed in each test image, although the final value of error which the model was reaching varied depending on the contrasts.

The test was conducted on several images to predict the visual scan path and the results were compared with the actual gaze data as shown below:

As seen in the above figures, our model is quite successful in estimating the saccadic scan path as it matches closely with the actual gaze data collected from eye tracker. However, as evident from Figures 6.4 and 6.6, our model generalizes the data and treats all regions that are not reconstructed properly to the same degree as of the same importance, which certainly is not the case. Our visual attention is also driven by high-level features such as faces and eyes, which this model is unable to predict. Hence, in the future, we can conduct experiments to determine the degree to which high-level features attract visual attention to build a model that would preserve the importance of high-level features.

FIGURE 6.8 Result example 1. (a) Predicted fixation points (b) Actual gaze data.

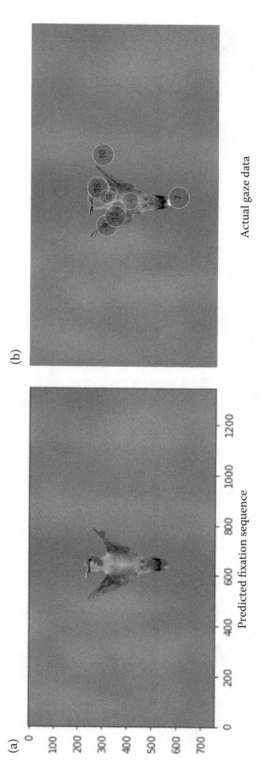

FIGURE 6.9   Result example 2.

REFERENCES

1. https://blog.keras.io/building-autoencoders-in-keras.html
2. Itti, L., Koch, C., Niebur, E. "A model of saliency-based visual attention for rapid scene analysis," *IEEE Trans. Pattern Anal. Mach. Intell.*, vol. 20, no. 11, pp. 1254–1259, Nov. 1998.
3. Xia, C., Han, J., Qi, F., Shi, G. "Predicting human saccadic scanpaths based on iterative representation learning," *IEEE Transactions on Image Processing*, vol. 28, no. 7, pp. 3502–3515, 2019.
4. Jain, S., Salau, A. O. "Detection of glaucoma using two dimensional tensor empirical wavelet transform," *SN Applied Sciences*, vol. 1, no. 11, p. 1417. DOI: 10.1007/s42452-019-1467-3
5. http://people.csail.mit.edu/tjudd/WherePeopleLook/index.html
6. https://colab.research.google.com/
7. https://keras.io/

7 Impact of IIOT in Future Industries
Opportunities and Challenges

G. Boopathi Raja

Assistant Professor, Department of ECE, Velalar College
of Engineering and Technology, Erode-638012

7.1 INTRODUCTION

In recent decades, advancements in wireless technology have led to the creation of a world-class technology, called the Internet of Things (IoT). The term Internet of Things was first coined by the British technology pioneer Kevin Ashton in the year 1998. The main idea was to assemble different objects and connect them to the Internet. IoT holds immense potential in several applications e.g. medical services, transportation, smart homes, and climate change. It has the capability to address challenges in businesses by achieving greater productivity, improved visibility, and cost control.

Various industrial developments have been made by the adoption of IIoT concept. It permits industries to collect and investigate huge amounts of data. It may help improve the standard presentation of modern structures by allowing management at different levels.

A number of advanced technology were named after IoT, such as Smart Manufacturing, Industry 4.0, and IIoT. The main objective of creating these advanced technologies was the utilization of cutting-edge innovations and applications e.g. 5G technology, Big Data analytics, Cloud computing, Machine Learning, Deep learning, Edge/Fog computing, and so on. In 2011, the German government started an initiative called "Industry 4.0", also termed as "Industrie 4.0." It was introduced to enhance the productivity with regards to assembling. This also helps trade and gather data associated with the entire lifecycle of any product.

IIoT is defined and stated in several forms, for instance, IIoT is defined as the network of smart and exceptionally grouped mechanical parts. It is used through continuous observation, efficient management, and regulation of modern cycles, properties, and operating time to conduct any process to meet optimum production rate with decreased costs.

DOI: 10.1201/9781003140443-7

IIoT, Internet of Underwater Things (IoUT), Internet of Drone Things (IoDT), and Internet of Medical Things (IoMT) are the subsets of IoT. In this, IIoT requires more significant levels of security as well as reliable communication without the disturbance of continuous modern activities due to strategic mechanical conditions. The main idea is to provide productive management of industrial resources and functions along with predictive support.

The vital contrasts between IoT and IIoT frameworks are described in Table 7.1. The fourth industrial revolution is based on the improvements in IIoT. This revolution is generally termed Industry 4.0. In this, it concentrates more on safety and effectiveness in assembling. The IIoT is expected to empower Industry 5.0 frameworks to limit the conflicts among humans and machines, [1]. Also, this will assist in accomplishing the significant personalization vision of the upcoming industrial revolution Industry 6.0. The essential movement in the field of IIoT represents the ongoing assessment of present industries. According to the latest report, by 2025, about 70 billion IoT-connected gadgets would be in use; however, by 2023, the share of IIoT in the worldwide market is anticipated to reach approximately USD 14.2 trillion.

Researchers discussed the trust-based communication challenges in IIoT [2], and scientists in [3] examined the guide to determine the network issues in remote IIoTs. Then again, analysts introduced a thorough survey of IIoT from some particular points of view. For example, energy-mindful information was steering in IIoTs [4], Cyber-Physical Systems (CPS) for IIoTs [5], and application sending methodologies in edge-figuring empowered IIoT frameworks [6], enormous information [7], and modernization of IIoT frameworks [8], [9], and [10]. Nonetheless, supposedly, we have given a modern outline of three significant regions in IIoT e.g. IIoT structures and systems, correspondence conventions, and information about the executive methods. We have covered the latest writing from

---

TABLE 7.1

Comparison of IIoT with IoT Framework

Parameter	Internet of Things (IoT)	Industrial IoT (IIoT)
Focus area	Smart gadgets	Industries
Application	General purpose	Industry oriented
Security and risk factors	Utilization based	Advanced and robust
Scalability	Smaller network	Larger network
Interoperability	Autonomous	Based on Cyber Physical System (CPS)
Accuracy level	Critically monitored	Synchronized with fractions of seconds
Response	Based on utilization and convenience	Operational efficient
Programmability	Off-site programming	On-site programming
Maintenance	Preference based on consumer	Properly pre-planned
Resilience	Not required	Fault tolerance needed

2015 to 2018. We have likewise proposed a more clear, straightforward meaning of IIoT and featured later difficulties looked at by the IIoT framework, [11], [12], and [13]. The latest empowering advances, which can assume a significant part in the achievement of IIoT frameworks, are additionally introduced. This investigation, moreover, featured the hole and center zones of AI in assembling, [14] and [15].

The remainder of this article is presented as follows: The cutting-edge research endeavors in IIoT are discussed in Section 2. The empowering advances of IIoT are briefly explained in Section 3. Finally, Section 4 describes the associated difficulties and issues in detail.

7.2 RESEARCH EFFORTS IN INDUSTRIAL IOT

The main objective of this section is to highlight the most recent research endeavors in the area of IIoT models and systems, corresponding conventions, and information about the executive procedures, [16], [17], and [18].

Figure 7.1 exhibits the architectural outline of IIoT. It consists of five domains, namely, business domains, application domains, information domains, operational domains, and control domains.

7.3 EMPOWERING TECHNOLOGIES FOR IIOT

Big Data, Cloud computing, Augmented Reality, Virtual Reality, Computer Vision, the Internet of Things, Artificial Intelligence, Machine Learning, Deep Learning, Human-to-Machine (H2M) Communication, and Machine-to-Machine (M2M) Communication form the pillars of IIoT, [19], [20], and [21]. Figure 7.2 shows the relation between Industry 4.0 and the advanced technologies.

7.3.1 INTERNET OF THINGS

IoT gadgets help in the collection of real-time information and actuation as in the case of any connected factory scenario. Being the essential element of IIoT, these gadgets monitor continuously the manufacturing plant resources across several countries. The entire cycle starts from crude material and terminates with finish items, [22]. It may be observed that by utilizing IoT gadgets the cost of labor resources and manual framework decreased. The IoT gadgets in an entirely associated IIoT framework are conveyed over all the manufacturing plant offices going from stockrooms to production areas and dissemination centers, [23], [24]. However, the arrangement, organization, checking, and support of these gadgets is a difficult assignment and requires an exceptionally qualified technician with expertise.

7.3.2 BLOCKCHAIN TECHNOLOGY

Blockchain technology plays a major role in bringing IIoT into reality, [25]. At present, much research is being done by various expert committees and industries

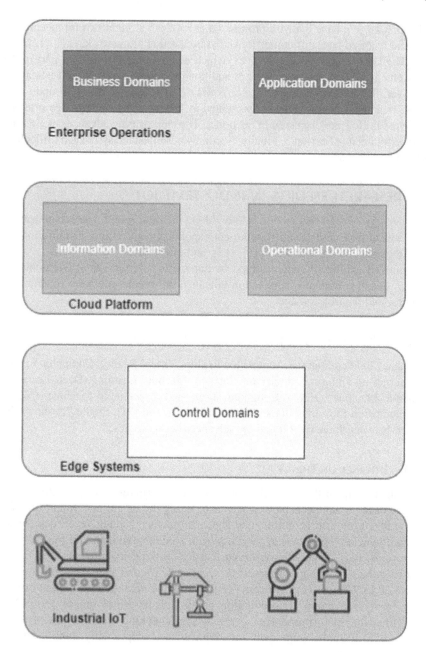

FIGURE 7.1 Architecture outline of Industrial IoT

on blockchain innovation in different fields e.g. medical services, supply chain, finance, and vehicle insurance.

The IoT-based devices employed in smart industries produce a large amount of data. The data received from the Internet-connected devices is multi-purposed.

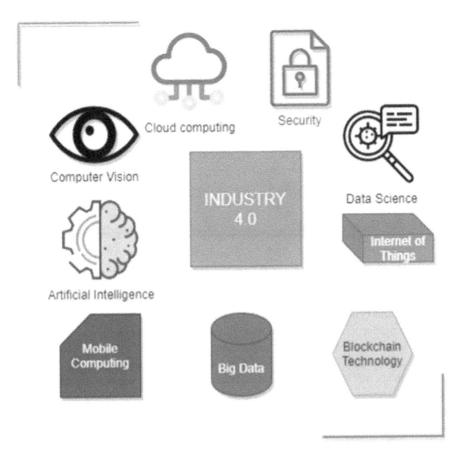

FIGURE 7.2 Relation between Industry 4.0 with advanced technologies

Hence, the obtained physical information is examined and processed for observing the performance of gadgets, peculiarity identification, monitoring of resources, predictive maintenance, and tracking the lifecycle of an entire product from raw material to a finished product.

In any event, it is difficult to deliver this vital data securely to all devices connected with the IIoT system. It is suitable for IIoT due to the special characteristics of blockchain technology, such as circulated existence, recognizability, altered opposition, survivability, confidence, protection, and intrinsic provenance of knowledge.

7.3.3 Cloud Computing

Radically distributed high-performance computing systems are required to process, calculate, interpret, and store data. In an IIoT system, distributed processing progressions offer figure, organization, and capability administration across all the offices. Backend fogs are directly interfaced with each and every associated device and application.

The models of cloud organization are organized as private (IIoT workers only had and supervised), public (outcast cloud merchants only asserted and directed), or creamer (whereby a mix of both assistance models is used). In this way, private cloud management models are not a fair option for recent participants or small- and medium-sized projects, as the development of server farms and the enlistment of qualified employees requires high investment.

Nevertheless, a large number of well-established companies worldwide are inclined toward arranging private cloud to guarantee well-being, security, and safety and adjust to modern dominant-handed undercover work.

7.3.4 Big Data Analytics

IIoT devices and systems account for a considerable proportion of data sources that are central to the significantly advanced prevailing figure frameworks for preparation and evaluation of huge data. Nonetheless, it is very testing to indicate when, how, and where to measure and investigate the large information with regard to idleness and genuine practicality in IIoT frameworks. To completely organize the large information examination administrations, IIoT frameworks empower various advances for huge information assortment, stockpiling, asset the executives, handling, dissecting, and activation. The information assortment provides connectors plenty of information sources, including sensors, keen gadgets, locally available information authorities, web-empowered information sources, and human-machine developments in IIoT frameworks, among others. Essentially, advancements in huge information stockpiling have encouraged installed, on reason, in-organization, and far off information stockpiling in cloud conditions. The information the board and preparing advances empower to deal with huge information close to the sensors, in edge workers, and in cloud server farms. The information investigation advances offer various instruments for information mining, AI, profound learning, and factual information examination at various layers within the IIoT frameworks. The innovations in action empower interactions between IIoT devices and their surrounding environments. Despite its multifaceted nature, enormous information handling and examination advancements have an essential function in cutting-edge IIoT frameworks.

7.3.5 Cyber Physical Systems (CPS) and Artificial Intelligence (AI)

Simulated advances in intelligence ensure that the IIoT system can function in a self-ruling and intelligent manner to reduce human interference and thereby enhance performance. By using nuanced AI progressions e.g. multi-expert systems and conversational AI, the AI innovations make IIoT autonomous. In addition, understanding is embedded in layers from sensors to devices to edge workers in IIoT systems, and by making assorted chase, progress, and desire estimates, cultivates cloud workers. IIoT systems allow undoubtedly advanced genuine mechanisms to limit human undertakings and intercessions, such as the development of systems and current robots.

CPS is embodied in built-in IoT gadgets that allow different sensors and actuators to function under mechanical conditions. In addition, these mounted embedded IoT gadgets facilitate informative knowledge planning for independent activities and upgrade competence in IIoT frameworks. In mechanical settings, these efficiencies range from various operational efficiencies to broad efficiencies in CPS and IIoT frameworks.

7.3.6 Augmented and Virtual Reality

Innovations in Augmented Reality (AR) help support automation professionals during complicated tasks e.g. gathering/de-amassing the apparatuses, complex modern items, and crucial frameworks. The AR advances empower to screen the laborers and machines during tasks and quickly produce modifies or notices to limit the blunders.

The developments of Virtual Reality (VR) enable the setup and rearrangement of modern capabilities and modules to be imagined before actual executions in IIoT frameworks. The use of VR helps mechanical plants and equipment to minimize re-arrangement times and chop off the shut-down holiday. The VR recreations are built by thinking about heterogeneity in CPS and IIoT frameworks by considering open guidelines.

7.4 CHALLENGES

A number of specialized problems have been brought together by the heterogeneous and complex life of IIoT systems, for example, interoperability, security, and protection, versatility, heterogeneity, dependability, and asset the executives. In any case, there are some difficulties that should have been settled. In this, we talk about these difficulties.

7.4.1 Efficient Management Procedure for Data

The wide range of reports about the heterogeneous IIoT devices allows the data volume to increase quickly. An increased proportion of high-speed information streams is made up of sensors and actuators that are connected to computer gadgets. The data obtained is removed from these separate IIoT gadgets, and the cloud staffs are used for future flexibility. Managing, transferring, accessing, and the power of information obtained is a daunting task and requires great effort.

With these difficulties, effective information about the board models is required. This information about the executive models is required to be prepared to proficiently deal with large amount of crude information generated by heterogeneous IIoT gadgets. These models ought to, similarly, furnish the information about the executive administrations with rapid information preparing, dependable and secure information stockpiling, recovery, and quick information stream.

7.4.2 HETEROGENEOUS IIoT SYSTEMS' COLLABORATIONS

An aggregation of a number of heterogeneous and multi-seller advances are the establishments of IIoT system e.g. mechanical machine, mechanical technology, IoT gadgets, sensors, actuators, passages, edge hubs, edge/cloud information workers (server farms), diverse wired/remote correspondence, and cell organizations (Wi-Fi, 5G). The reconciliation and joint efforts between these heterogeneous and multi-seller advances based on the IIoT method is a tricky challenge. The teamwork and joint effort are all further tested by various components, such as synchronization, asset sharing, knowledge sharing, interoperability, and information security. More innovative work endeavors are as yet needed for adaptable and productive methods for coordinated effort and interoperability.

7.4.3 ROBUST AND SCALABLE ANALYTICAL TECHNIQUES FOR BIG DATA

There is demand for powerful and adaptable broad information exam developments to achieve the vision of IIoT and to get maximum benefits from the high volume of information provided by IIoT gadgets. The traditional information base administration frameworks cannot create the ideal outcomes, as these frameworks cannot measure and dissect the high extent of information effectively. Preparing the IIoT information progressively a basic errand as this information is utilized for basic real-time mechanical robotization activities e.g. foresee a glitch, prescient upkeep, increment creation, and diminish down-times and oddity recognition.

To meet the fluctuating needs of IIoT applications (e.g. traffic rates, dormancy, and reliability, etc.), the powerful and efficient handling of IIoT gadget information requires successful and ongoing large-scale information investigation innovations.

Such information analysis advances often include information planning and representations to support the entire lifecycle of the product (e.g. development, testing, customer reviews, and after-deal administrations, etc.) to achieve a complete understanding of business.

7.4.4 TRUST IN INDUSTRIAL IoT-BASED FRAMEWORKS

Purchaser acknowledgment and transformation are straightforwardly associated with the accomplishment of any innovation and are exceptionally impacted by consumer trust in these advances. The effective arrangement of IIoT-based frameworks by business clients (e.g. proprietors of the specific business) is likewise influenced by the trust in these IIoT frameworks. IIoT frameworks are at their outset and the greater part of the ongoing exploration writing featured security and protection as a significant test looked into by these frameworks. Security and protection of innovation are firmly connected with the trust of their clients, in this manner, frail security and protection of IIoT frameworks will deter the clients from appropriations of these IIoT frameworks. Consequently, for fruitful sending and transformation of IIoT frameworks in the business, client trust must be arranged appropriately with compelling client trust models. Thus, more examination in the region of client trust models is needed for the effective acknowledgment of IIoT frameworks.

7.4.5 Wireless Technology and Protocols Coexistence with IIoT

Lately, the IIoT is attracting development considerations from both the scholarly world and the industry. Correspondence in IIoT is required for the trading of data. Hence, correspondence in IIoT must have the option to interface a large number of heterogeneous gadgets, give enough transmission capacity to move information, and offer deterministic conduct with low inactivity.

Some modern applications have timeless, unstable quality, accessibility, and security requirements. Numerous communication developments, meetings, and procedures have been implemented in IIoT. Late time, remote control (WLAN, IEEE 802.15 (WSN)), and remote gadgets receive more consideration than previous cable failures. Remote communication also leads to many problems. The integration of various remote structures and meetings is an important test for IIoT. Investigation is what creates the literature and the meeting is the best application. A new design or assembly cannot offer all of the highlights and features that match the unique needs of the app in IIoT. In this way, the determination of new texts and textbooks is a major test.

7.4.6 Enabling Decentralization on the Edge

The information sources and huge production of persistent information streams lead to heterogeneity in the network. Hence, it requires accessibility in performing some operations by doing desired computation, networking, and memory elements on the edge of the Internet. Edge computing empowers to advance the end nodes, in any case, the edge benefits just as information the board on the edge is completely arranged through incorporated cloud regulators. This reliance not just improves the prerequisites of highly accessible communication channels, it also leads to a single point of failure in the modern industrial framework.

The decentralization concept in end-to-end cloud management can help mitigate the aforementioned issues. The integration of end servers with blockchain technologies may provide a consistent inventory of asset requirements from the latest apps and gadgets. Also, the servers can enable delegated power supply and management integration without relying on integrated controllers.

7.4.7 Development of IoT-Specific Operating Systems

The operating system (OS) used in the IoT aims to provide services only within the core specifications of IoT gadgets based on the limitations of memory, size, capacity, and limit setting. IoT OS is well known for efficient handling of memory, versatile support with a programming language, scheduling, and design, and is tested in [26]. Through a test network, TinyOS and Contiki are commonly preferred IoT OSs as they meet most of the requirements needed for any application. A limited memory, constant efficiency, power, resource functions, protection, scheduling, reliable communication, information storage, end device management, and hardware agnostic operations must be the key requirements of the IoT OS.

For any IoT-based applications e.g. smart healthcare, smart grid, smart home, and automated ticket counter, the applications in IIoT require more rigid QoS necessities. Reliable communication, privacy and security, power utilization, interoperability, supporting in heterogeneous gadgets, and transmission capacity utilization are the vital difficulties in IIoT. In [27], for IoT and its applications, researchers provide an in-depth analysis of the most critical key highlights of an OS. In addition, they analyzed IoT OS for connectivity, difficulties, and contextual studies as per their executed inventions.

7.4.8 PUBLIC SAFETY IN IIoT

IIoT-based social welfare schemes involved during the occurrences of risk and disaster situations must be specially organized. In the unlikely event with regard to the current state of affairs, the welfare of trained staff and gear depends on good recognition of occasions, years of goodwill, site closure, and warning notices to co-ops problem specialists e.g. local fire team soldiers, rescue vehicle, command unit disaster, traffic police, etc., organizations that need the law. In addition to this, the invisibility or breach of the literature in unfortunate regions leads to a more complicated problem. Also, the communication and collaboration among different IIoT gadgets and other linking platforms cause an exploratory problem.

Numerous ongoing examinations handle proposed network structures that depend on efficient interchanges for IoT and future savvy urban areas by using Unmanned Air Vehicles, SONET Digital Networks (SDN), Edge figuring and other progressed communication advancements, such as 4G, 5G, and so on. Nonetheless, research endeavors are expected to plan catastrophe tough, independent structures that would empower profoundly proficient correspondence in ordinary conditions. Likewise, these designs are expected to guarantee calamity recuperation components for public security if there should be an occurrence of any crisis.

7.5 CONCLUSION

In the industrial sector, the aim of IIoT-based frameworks is to collect and disseminate a wide variety of data that can be used, updated, and expanded to provide new ways of management with the traditional presentation of frameworks. In this chapter, the latest cutting-edge research efforts on IIoT are highlighted. The various empowerment methods identified by IIoT are also introduced. An important open-ended testing challenge is isolated from the fruitful IIoT system.

These tests result in the failure of the IIoT to prevent the various difficulties shown in this investigation, which include effective board-based data, integrated efforts within the incomparable IIoT framework, robust and flexible data on logical development, reliance on IIoT frameworks, and integration of new remote meetings in IIoT.

REFERENCES

1. W. Z. Khan, M. H. Rehman, H. M. Zangoti, M. K. Afzal, N. Armi, K. Salah, Industrial internet of things: Recent advances, enabling technologies and open challenges, *Computers & Electrical Engineering* 2020.
2. C. Zhu, J. J. P. C. Rodrigues, V. C. M. Leung, L. Shu, L. T. Yang, Trust-based communication for the industrial internet of things, *IEEE Communications Magazine* 56 (2) (2018) 16–22 (Feb 2018). doi:10.1109/MCOM.2018.1700592.
3. S. Mumtaz, A. Alsohaily, Z. Pang, A. Rayes, K. F. Tsang, J. Rodriguez, Massive internet of things for industrial applications: Addressing wireless iiot connectivity challenges and ecosystem fragmentation, *IEEE Industrial Electronics Magazine* 11 (1) (2017) 28–33 (2017).
4. N. B. Long, H. Tran-Dang, D. Kim, Energy-aware real-time routing for large-scale industrial internet of things, *IEEE Internet of Things Journal* 5 (3) (2018) 2190–2199 (June 2018). doi:10.1109/JIOT.2018.2827050.
5. H. Xu, W. Yu, D. Gri_th, N. Golmie, A survey on industrial internet of things: A cyber-physical systems perspective, *IEEE Access* 6 (2018) 78238–78259 (2018). doi:10.1109/ACCESS.2018.2884906.
6. M. Aazam, S. Zeadally, K. A. Harras, Deploying fog computing in industrial internet of things and industry 4.0, *IEEE Transactions on Industrial Informatics* 14 (10) (2018) 4674–4682 (Oct 2018). doi:10.1109/TII.2018.2855198.
7. K. Al-Gumaei, K. Schuba, A. Friesen, S. Heymann, C. Pieper, F. Pethig, S. Schriegel, A survey of internet of things and big data integrated solutions for industrie 4.0, in: 2018 IEEE 23rd International Conference on Emerging Technologies and Factory Automation (ETFA), Vol. 1, IEEE, 2018, pp. 1417–1424 (2018).
8. C. Perera, C. H. Liu, S. Jayawardena, The emerging internet of things marketplace from an industrial perspective: A survey, *IEEE Transactions on Emerging Topics in Computing* 3 (4) (2015) 585–598 (Dec 2015). doi:10.1109/TETC.2015.23 90034.
9. E. Sisinni, A. Saifullah, S. Han, U. Jennehag, M. Gidlund, Industrial internet of things: Challenges, opportunities, and directions, *IEEE Transactions on Industrial Informatics* 14 (11) (2018) 4724–4734 (Nov 2018). doi:10.1109/TII.2 018.2852491.
10. S. Jeschke, C. Brecher, T. Meisen, D. Ozdemir, T. Eschert, Industrial internet of things and cyber manufacturing systems, in: *Industrial Internet of Things*, Springer, 2017, pp. 3–19 (2017).
11. Y. Liao, E. de Freitas Rocha Loures, F. Deschamps, Industrial internet of things: A systematic literature review and insights, *IEEE Internet of Things Journal* 5 (6) (2018) 4515–4525 (Dec 2018). doi:10.1109/JIOT.2018.2834151.
12. S. Lin, B. Miller, J. Durand, G. Bleakley, A. Chigani, R. Martin, B. Murphy, M. Crawford, The industrial internet of things volume g1: reference architecture, *Industrial Internet Consortium* 5 (2017) 10–46 (2017).
13. G. Campobello, M. Castano, A. Fucile, A. Segreto, Weva: A complete solution for industrial internet of things, in: International Conference on Ad-Hoc Networks and Wireless, Springer, 2017, pp. 231–238 (2017).
14. C. K. M. Lee, S. Z. Zhang, K. K. H. Ng, Development of an industrial internet of things suite for smart factory towards re-industrialization, *Advances in Manufacturing* 5 (4) (2017) 335–343 (nov 2017). doi:10.1007/s40436-017-0197-2.
15. W. Z. Khan, M. Y. Aalsalem, M. K. Khan, M. S. Hossain, M. Atiquzzaman, A reliable internet of things based architecture for oil and gas industry, in: Advanced Communication Technology (ICACT), 2017 19th International Conference on, IEEE, 2017, pp. 705–710 (2017).

16. F. Tao, J. Cheng, Q. Qi, IIHub: An industrial internet-of-things hub toward smart manufacturing based on cyber-physical system, *IEEE Transactions on Industrial Informatics* 14 (5) (2018) 2271–2280 (may 2018). doi:10.1109/tii.2017.2759178.
17. B. Martinez, X. Vilajosana, I. Kim, J. Zhou, P. Tuset-Peir_o, A. Xhafa, D. Poissonnier, X. Lu, I3mote: An open development platform for the intelligent industrial internet, *Sensors* 17 (5) (2017) 986 (apr 2017). doi:10.3390/s17050986.
18. Z. Kaleem, M. Yousaf, A. Qamar, A. Ahmad, T. Q. Duong, W. Choi, A. Jamalipour, Uav-empowered disaster-resilient edge architecture for delay-sensitive communication, *IEEE Network* (2019) 1–9 (2019). doi:10.1109/MNET.2019.1800431.
19. Z. Meng, Z..Wu, C. Muvianto, J. Gray, A data-oriented m2m messaging mechanism for industrial iot applications, *IEEE Internet of Things Journal* 4 (1) (2017) 236–246 (2017).
20. W. Yang, Y. Wan, Q. Wang, Enhanced secure time synchronisation protocol for ieee802. 15.4 e-based industrial internet of things, *IET Information Security* 11 (6) (2017) 369–376 (2017).
21. T. Qiu, Y. Zhang, D. Qiao, X. Zhang, M. L. Wymore, A. K. Sangaiah, A robust time synchronization scheme for industrial internet of things, *IEEE Transactions on Industrial Informatics* 14 (8) (2017) 3570–3580 (2017).
22. S. Rao, R. Shorey, E_cient device-to-device association and data aggregation in industrial iot systems, in: Communication Systems and Networks (COMSNETS), 2017 9th International Conference on, IEEE, 2017, pp. 314–321 (2017).
23. T. P. Raptis, A. Passarella, M. Conti, Maximizing industrial iot network lifetime under latency constraints through edge data distribution, in: 1st IEEE International Conference on Industrial Cyber-Physical Systems, (ICPS) (May 2018), available at http://cnd.iit. cnr. it/traptis/2018-raptis-icps.pdf.
24. M. H. ur Rehman, I. Yaqoob, K. Salah, M. Imran, P. P. Jayaraman, C. Perera, The role of big data analytics in industrial internet of things, *Future Generation Computer Systems* 99 (2019) 247–259 (April 2019). doi:10.1016/j.future.2019.04.020.
25. D. Miller, Blockchain and the internet of things in the industrial sector, *IT Professional* 20 (3) (2018) 15–18 (2018).
26. F. Javed, M. K. Afzal, M. Sharif, B.-S. Kim, Internet of things (iot) operating systems support, networking technologies, applications, and challenges: A comparative review, *IEEE Communications Surveys & Tutorials* 20 (3) (2018) 2062–2100 (2018).
27. Y. B. Zikria, S. W. Kim, O. Hahm, M. K. Afzal, M. Y. Aalsalem, Internet of things (iot) operating systems management: Opportunities, challenges, and solution, *Sensors* (2019) 1–10 (2019).

8 Privacy and Ethical Issues in Digitalization World

Renu Bala
Research Scholar, Chaudhary Devi Lal University,
Sirsa (Haryana)

8.1 INTRODUCTION

Privacy, security, and trust are the three terms interlinked with each other in a similar way as law and ethics are related to each other. The term data privacy represents how the data should be collected, used, and accessed by considering legal rights (*Lee et al., 2016*; *Salau et al., 2021*). On the other side, ethics denotes responsibilities that sometimes become an obligation to pursue (*Knoppers & Thorogood, 2017*). At present, in this technology-driven era, Internet-based research is vital, and most researchers follow this method widely to fulfill the purpose of data collection and analysis. The umbrella of Internet-based research covers video conferencing for interviews, online surveys, analysis of "e-conversations," web page content analysis, discussion blogs, chat rooms, and email, etc. *(Cox, 2012)*. With the passage of time, many social and technological changes can be noticed, and these changes are altering the scale and scope of the data for researchers *(Fiesler)*. In addition to the many advantages of the latest technology, some drawbacks also exist. Data sharing and data storage provide different kinds of benefits to researchers, but it creates many hurdles also from the security perspective of data, and in this present scenario, traditional privacy mechanism and security system are not adequately sufficient to cope with data explosion. As a result, researchers have to face different kinds of data issues during their original work, such as the modification of secondary data, storing of data, sharing of data, etc., and, in these situations, ethical decision-making occurs (*Boyd et al., 2016*). Due to the advanced technology, the chances of misuse of data are increasing day by day, due to this which it becomes tough to follow ethics (*Hand, 2018*). Researchers must face various types of data privacy challenges during their work, such as securing the computations under distributed programming frameworks, securing data storage and the transactional logs, data provenance, end-point filtration and validation, and real-time security monitoring (*Victor & Lopez, 2018*), etc. So, the government and other private institutions should focus on developing different kinds of new software under the latest technology, which can protect data by maintaining its privacy and avoiding misuse, modification, and wrong interpretation, etc.

DOI: 10.1201/9781003140443-8

8.2 LITERATURE REVIEW

Nigam et al. highlighted the significance of the digital world in different streams. The significant role of digitalization in the IT sector and research field has been defined in the study. *Soni & Pandey* described digitalization in e-marketing and explained how consumers benefit from digitalized services. *Neumeier* explained that digitalization is playing a positive role in every sector of the economy. *Kumar et al.* highlighted the role of technology in the retail sector and concluded that technology had created very significant outcomes in retailing services. *Shallu et al.* highlighted the effects of digitalization on India's economy and mentioned the challenges faced by India's government to solve various technological issues. *Nadkarni & Prugl* presented the opportunities in the world of digitalization and the ways to exploit them.

8.3 CHALLENGES IN THE STREAM OF DATA PRIVACY

Table 8.1 shows that there exist different kinds of challenges in data privacy. These highlighted challenges include sharing and publishing network traces, collection of network traces, data encryption, access controls, server consolidation, protection from insider threats, data trustworthiness, reconcile data security and privacy, securing the computations under distributed programming frameworks, securing data storage and transactional logs, data provenance, end-point filtration and validation, real-time secure monitoring, input validating and filtration, granular accessed control, unsecured data storage, cryptographically enforce data-centric security, scalable privacy-preserving analytics and data

TABLE 8.1

Challenges in the Stream of Data Privacy

S. No.	Challenges in Data Privacy	S. No.	Challenges in Data Privacy
1.	Share and publish network traces	11.	Data provenance
2.	Collection of network traces	12.	Unsecured data storage
3.	Data encryption	13.	Real-time secure monitoring
4.	Access controls	14.	Input validating and filtration
5.	Server consolidation	15.	Granular accessed control
6.	Protection from the insider threats	16.	End-point filtration and validation
7.	Data trustworthiness	17.	Cryptographically enforce data-centric security
8.	Reconcile data security and privacy	18.	Scalable privacy-preserving analytics and data mining
9.	Securing computed data under the distributed programming frameworks	19.	Repetition of scraping by multiple accounts
10.	Securing data storage and the transactional logs	20.	Researcher misrepresentation as well as fake profile data

TABLE 8.2
Ethical Issues in Data Privacy

S. No.	Ethical Issues	S. No.	Ethical Issues
1.	Misuse of social licensing	6.	Negative exploitation of research tools, techniques, and standards, etc.
2.	Make a balance between risks and benefits	7.	Misuse of data
3.	Wrong use of "right to science" in international law	8.	Break the privacy breach
4.	Wrong governance of Big Data	9.	Wrong interpretation of data as per requirements
5.	Modification of cybersecurity research ethics	10.	Consider the research as a source of money only

mining, repetition of scraping by multiple accounts, researcher misrepresentation, as well as fake profile data, and much more.

8.4 ETHICAL ISSUES IN DATA PRIVACY

Table 8.2 shows that there exist several types of ethical issues in data privacy, and the major issues include the misuse of social licensing, making a balance between risks and benefits, inappropriate use of "right to science" in the international law, bad governance of Big Data, the modification of cybersecurity research ethics, negative exploitation of research tools, techniques, and standards, etc., misuse of data, break the privacy breach, wrong interpretation of data as per requirements, consider the research as a source of money only, etc.

8.5 CONCLUSION

The existing challenges in data security ruin the actual meaning of data and lose its authenticity. Researchers, who make serious and hard efforts in their work, feel demoralized due to theft and data misuse. Actually, data privacy has become a major issue in this era, and to fight against such types of issues, private and government organizations are paying attention and making good efforts to introduce new kinds of techniques and software, and taking other measures for data security. However, strict rules and punishment should be fixed for people who exploit original work and misuse data.

This study presents an organized and detailed review of empirical studies in data privacy challenges and ethical issues. It references the researchers to carry out their research in this respective field and provide a clear picture of data privacy challenges. After analyzing the abovementioned challenges and ethical issues, this study also provides knowledge to researchers regarding the present challenges in the respective fields. After analyzing difficulties in this area, solutions can be found to solve problems and overcome challenges.

REFERENCES

Abacahin, J. & Amil, D. *et al.* (2012). Sales and Inventory System of Edmar Marketing. *Advancing Information Technology Research*, 3(1), 259–282, ISSN: 2094-9626.

Abouelmehdi, K. & Hessane, A. *et al.* (2018). Big Healthcare Data: Preserving Security and Privacy. *Journal of Big Data*, 5(1), 1–18, ISSN: 21961115.

Agapay, R. S. & Babanto, L. S. *et al.* (2012). Online Book Information System of Liceo de Cagayan University. *Advancing Information Technology Research*, 3(1), ISSN: 2094-9626.

Ahmad, S. & Azam, M. *et al.* (2018). A Survey on Security and Privacy of Big Data. *International Journal of Engineering Research in Computer Science and Engineering (IJERCSE)*, 5(4), 96–100, ISSN: 2394-2320.

Alnajrani, H. M. & Norman, A. *et al.* (2020). Privacy and Data Protection in Mobile Cloud Computing: A Systematic Mapping Study. *PLOS ONE*, 15(6), 1–28, ISSN: 1932-6203.

Andersen, A. & Saus, M. (2017). Privacy-Preserving Distributed Computation of Community Health Research Data, *Procedia Computer Science*, 113(1), 633–640, ISSN: 00002010.

Angeles, B. D. & Rado, L. M. (2012). LAN-Based Management Information System of Petron Gas Service Station - Puerto Branch. *Advancing Information Technology Research*, 3(1), ISSN: 2094-9626.

Bao, R. & Chen, Z. *et al.* (2017). Challenges and Techniques in Big Data Security and Privacy: A Review. *WILEY*, 1(4), 1–8, ISSN: 2048-416X.

Bertino, E. (2014). Data Security – Challenges and Research Opportunities. Paper presented at Workshop on Secure Data Management, 1–5.

Bhandari, R. & Hans, V. *et al.* (2016). Big Data Security – Challenges and Recommendations. *International Journal of Computer Sciences and Engineering*, 4(1), 93–98, ISSN: 2347-2693.

Burstein, A. J. (2008). Conducting Cyber Security Research Legally and Ethically. Paper presented at First USENIX Workshop on Large-Scale Exploits and Emergent Threats, San Francisco, CA, Proceedings.

Buted, D. R. & Gillespie, N. S. *et al.* (2014). Effects of Social Media in the Tourism Industry of Batangas Province. *Asia Pacific Journal of Multidisciplinary Research*, 2(3), 123–131, ISSN 2350-8442.

Boyd, D. & Keller, E. F. *et al.* (2016). Supporting Ethical Data Research: An Exploratory Study of Emerging Issues in Big Data and Technical Research. *Data & Society*, 1–26.

Chiauzzi, E. & Wicks, P. (2019). Digital Trespass: Ethical and Terms-of-Use Violations by Researchers Accessing Data from an Online Patient Community. *Journal of Medical Internet Research*, 21(2), 1–12, ISSN: 1438-8871.

Cooke, L. (2018). Privacy, Libraries and the Era of Big Data. *IFLA (International Federation of Library Associations and Institutions) Journal*, 44(3), 167–169, ISSN: 1745-2651.

Cox, D. (2012). A Review of Research Ethics in Internet-Based Research. *Practitioner Research in Higher Education*, 6(1), 50–57, ISSN: 1755-1382.

Fabiano, N. (2019). Ethics and the Protection of Personal Data. *Systemics, Cybernetics and Informatics*, 17(2), 58–64, ISSN: 1690-4524.

Floridi, L. & Taddeo, M. (2016). What Is Data Ethics? *Philosophical Transactions of the Royal Society*, 374(2083), 1–8, ISSN: 2053-9223.

Hand, D. (2018). Aspects of Data Ethics in a Changing World: Where Are We Now? *Big Data*, 6(3), 176–190, ISSN: 2214-5796.

Hasselbalch, G. (2019). Making Sense of Data Ethics. The Powers Behind the Data Ethics Debate in European Policymaking, *Internet Policy Review*, 8(2), 1–19, ISSN: 2197-6775

Inukollu, V. N. & Arsi, S. *et al.* (2014). Security Issues Associated with Big Data in Cloud Computing. *International Journal of Network Security & its Applications (IJNSA)*, 6(3), 45–56, ISSN: 0974-9330.

Kaiser, K. (2009). Protecting Respondent Confidentiality in Qualitative Research. *Qualitative Health Research*, 19(11), 1632–1641, ISSN: 1049-7323.

Kantarcioglu, M. & Ferrari, E. (2019). Research Challenges at the Intersection of Big Data, Security and Privacy. *Frontiers in Big Data*, 2(1), 1–6, ISSN: 2624-909X.

Kante, M. (2017). A Review of Big Data Security and Privacy Issues. *Mara International Journal of Scientific & Research Publications*, 1(1), 49–54, ISSN: 2523-1456.

Kenneally, E. & Bailey, M. (2014). Cyber-Security Research Ethics Dialogue & Strategy Workshop. *Computer Communication Review*, 44(2), 76–79, ISSN: 01464833.

Knoppers, B. M. & Thorogood, A. M. (2017). Ethics and Big Data in Health. *Current Opinion in Systems Biology*, 4(1), 53–57, ISSN: 24523100.

Lane, J. & Schur, C. (2020). Balancing Access to Health Data and Privacy: A Review of the Issues and Approaches for the Future. *Health Services Research*, 45(5), 1456–1467, ISSN: 00179124.

Lee, W. & Zankl, W. *et al.* (2016). An Ethical Approach to Data Privacy Protection. *ISACA Journal*, 6(1), 1–9, ISSN: 19441967.

Magno, K. J. H. & Pabico, J. P. (2014). Digital Anthropometry: Model, Implementation, and Application. *Asia Pacific Journal of Multidisciplinary Research*, 2(3), 82–88, ISSN: 2350-8442.

Mehmood, A. & Natgunanathan, I. et al. (2016). Protection of Big Data Privacy. IEEE, 4(1), 1821–1834, ISSN: 2169-3536.

Mikalef, P. & Boura, M. et al. (2019). Big Data Analytics and Firm Performance: Findings from a Mixed-Method Approach. *Journal of Business Research*, 98 (1), 261–276, ISSN: 0148-2963.

Nuyda, D. & Calma, J. R. *et al.* (2012). Online Inventory and Monitoring System of Cagayan de Oro City Health Office. *Advancing Information Technology Research*, 3(1), 69–110, ISSN: 2094-9626.

Padhy, R. P. & Patra, M. R. *et al.* (2011). Cloud Computing: Security Issues and Research Challenges. *International Journal of Computer Science and Information Technology & Security (IJCSITS)*, 1(2), 136–146.

Salau, A. O., Marriwala, N., & Athaee, M. (2021). Data Security in Wireless Sensor Networks: Attacks and Countermeasures. *Lecture Notes in Networks and Systems*, vol. 140. Springer, Singapore.

Victor, N. & Lopez, D. (2018). Privacy preserving big data publishing: challenges, techniques, and architectures. In *HCI Challenges and Privacy Preservation in Big Data Security* (pp. 47–70). IGI Global.

9 A Review on Smart Traffic Management System

Tarun Jaiswal, Manju Pandey, and Priyanka Tripathi
Department of Computer Applications, National Institute of Technology (NIT), Raipur (C.G.), INDIA

9.1 INTRODUCTION

Traffic supervision is one of the major set-up track events confronted as a result of rising developments in nations nowadays. Industrialized countries and smart cities are now accumulated via IoT and to their improvement to reduce problems associated with the traffic signal. The values of the vehicle have been cultured quickly between humans in each kind of countries. In large cities, it is communal designed for the public to like better equine their personal vehicles no substance how worthy or depraved the communal conveyance is or allowing for how considerable time and cost is it going to take used for them to influence their endpoint.

Through the realm, there is a constantly growing number of cars [1] that chief to the everyday experiment of discovery a parking space in metropolitan zones. Between numerous categories of overcrowding comparable to the standard flow overcrowding, there exist correspondingly numerous forms of parking-related overcrowding [2].

Some of the adverse effects of traffic congestion are environmental issues, air pollution, energy consumption, parking space scarcity, and traffic-related expenses of around hundreds of billions of dollars each year [3].

The authors [4] presented, the growths of technology and conveys grasps accumulated the prevailing by the supreme degree of improvement. The study demonstrated the challenges on the maneuver, enactment policy and succeeding ideas of smart cities in India [5].

Smart traffic supervision is a classification castoff to standardize city traffic. It customs sensors and traffic signals to observation, control and feedback to traffic environments. These decentralized accomplished sensors and traffic signals are established on the city's leading routes. The significant objective of smart traffic supervision structures includes: improving traffic conditions and reduces the congestion over the road; providing priority for emergency vehicles in real-time according to the traffic conditions; lowering the emission of the

DOI: 10.1201/9781003140443-9

pollute net gas; setting the ordering value for the traffic data, high preference set the emergency vehicle likewise ambulance throughout cities; creating a more effective traffic management system for monitoring and managing traffic incidences. Still, many metro cities depend on the conventional traffic light/ signals for traffic management, but the new IoT-based traffic management system gains its input data from the real-time environments and regulates traffic signals according to the traffic demands. The IoT-based smart traffic control system consists of generally three devices, namely: a central Hub; smart traffic signals; counters, and cameras. The counter and the camera provide the real-time necessary traffic condition to the system, the smart traffic system accordingly changes the motion of the traffic light to maintain the smooth flow of traffic. The traffic is smooth by means of lowering the no of que on the road, providing the safe lane intersection, scheduling the route of heavy vehicles like buses and truck etc. thereby it enhances the safety and efficiency of traffic and also reduces the venomous gaze emitted from the vehicles. IoT-based smart traffic signals/lights are very useful for tackling the traffic jam condition or safely crossing the vehicle at the intersection lane, this smart light is smart enough that it can detect the traffic pattern and accordingly manipulate the signal or light over the signals.

9.2 DECREASE OF TRAFFIC MOVEMENT THROUGH THE SMART SIGNALS

Detecting overcrowding: Smart traffic identify the traffic in the real-time by means of numerous sensors placed over the infrastructure or vehicle and smart traffic management system identify density of the traffic and if it sense the heavy traffic over some lane or road then it divert it; Coordinating commotion among traffic lights: For synchronization of traffic light sensors are placed at various location that monitor the traffic and maintain the synchronization between the traffic signals; Apprising traffic light interval in real-time: The traffic light works on real-time scenario, and are not permitted to deliver signals at predefined time interludes; Apprising and notifying drivers of superlative speeds: Some traffic lights are smart enough that they inform the driver about the next traffic light status and if driver maintain the steady and correct speed as suggested by the system then driver find the next signal clear; Arranging transportation movement: Traffic signals/lights arranged public conveyance over reserved vehicles e.g. ambulance, taxis, cycle, buses etc. continuously obtain predilection over private vehicle in the intersecting route.

9.3 ROLE OF IOT AND BIG DATA IN TRAFFIC MANAGEMENT

The smart traffic management system controls all operations from Big Data by IoT. In the perspective of traffic administration, IoT mentioned smart, sensors enabled devices, autonomous vehicle systems, and smart mobile devices. The device gathers the data and transfers it to the centralized server system for examinations. Big Data is accountable for examining data and used to enhance traffic density and flow. In smart cities, integrating IoT expedients as an arrangement of sensors and detectors bounded everywhere in cities. IoT and Big Data influence the traffic controlling arrangement by a smart traffic light, smart parking, smart roads, smart data analytics, and public transportation, etc.

Smart traffic lights: inbuilt IoT sensor fixed on roads and vehicles in a specific position thru this accumulated the information in a centralized system. The objective of the smart traffic light that is inbuilt with IoT and Big Data provides the optimal solution. *Smart parking*: IoT-based GPS navigation system to detect and provide parking space via smart mobile devices.

Smart roads: Road sensors detect the location, alcoholic driver position, and speed of the vehicle under law enforcement conditions. Inbuilt sensor warns and alerts for accident situation.

Smart-data analytics: AI-based s/w and simulation used to reduce the traffic congestion for this IoT enabled visualization facility to gather the material from locations and access points. Observation of visualization frames to detect the person who responsible for accidents and vulnerabilities.

Public transportation: Through smart mobile devices and inbuilt IoT sensors, a person quickly finds out the public transport timing and locations. However, the weather conditioned and some delay conditioned affected the normal public transportation system. So real-time autonomous public transport system resolve issues.

9.4 ADVANTAGES AND DISADVANTAGES OF TRAFFIC MANAGEMENT SYSTEM

9.4.1 ADVANTAGES

Communication: Communication among devices is encouraged by IoT, also known as Machine-to-machine (M2M) communication. As a result, several physical machines may remain connected, and, therefore, the overall clarity is accessible with lower inefficiencies.

Automation and Control: There is a significant amount of automation and monitoring in the workings due to physical objects getting remotely and centrally linked and managed via wireless networks. The systems are able to interact with each other without human involvement, leading to faster and timely production.

Information: It is clear that having more details allows you to make smarter choices. Whether it's mundane choices like knowing what to buy at the grocery store or getting sufficient widgets and supplies for your company, information is power, and more knowledge is better.

Monitor: Surveillance is the second most apparent benefit of the IoT. Furthermore, understanding the precise amount of materials or the air conditioner in your home will provide more details that may not have been conveniently gathered previously. Realizing that you are short on milk or printer ink, for example, might save you another ride in the near future to the supermarket. In addition, recording the expiration of commodities will increase protection.

Time: The length of time saved due to IoT may be very high, as shown in the earlier cases. And we could all use more time in today's modern life.

Money: The IoT's most significant benefit is saving money. The Internet of Things would be very commonly embraced if the price of the marking and tracking devices is smaller than the amount of money saved. In their everyday lives, IoT essentially proves to be very useful to individuals by making the machines interact efficiently with each other, protecting and preserving resources and costs. It allows our systems to be effective by allowing the data to be exchanged and transmitted among systems and then converting it in our desired manner.

Efficient and Saves Time: Better performance is given by machine-to-machine interaction, so reliable data can be achieved easily. Saving precious time helps in this. It encourages individuals to do more innovative jobs instead of performing the same jobs each day.

Better Quality of Life: Many of this technology's implementations culminate in improved comfort, accessibility, and efficient governance, thereby enhancing the quality of life.

9.4.2 Disadvantages Compatibility

There is no international compatibility requirement for tagging and tracking devices at present. This drawback, I assume, is the toughest to solve. These device manufacturing firms only need to agree to a specification, such as Bluetooth, USB, etc. This doesn't need something new or creative.

Complexity: There are more risks of loss than in all dynamic systems. Failures could skyrocket for the Internet of Things. Let's assume, for example, that you and your wife both get a message that says that your milk has expired, and you both stop at a store on your way home, and you both buy milk. As a consequence, you and your partner have bought double the quantity you would like. Or maybe a program error would automatically end up buying a new ink cartridge for your printer every hour for a couple of days, or at least after each power outage when you only need a single substitute.

Privacy/Security: The chance of losing privacy rises with all of this IoT information being shared. For example, how well will the data be maintained and transmitted with encryption? Would you want to hear what drugs you are taking or your financial status from your neighbours or employers? Safety: Since all household equipment, heavy machines, public sector utilities such as water and transportation, and many other gadgets are all linked to the Internet, there is a great deal of knowledge available. Such data is vulnerable to attack by hackers. If unwanted intruders obtained personal and classified information, it would be rather catastrophic.

Lesser Employment of Manpower: In the wake of the automation of everyday operations, unskilled laborers, and peoples can end up losing their jobs. This can lead to problems with huge unemployment in the city. This is a concern since the introduction of any advancement which can be addressed with education. Generally, with everyday operations being streamlined, there would be fewer expectations for human resources, especially employees and less trained employees. This will establish the problem of unemployment in the community.

9.5 LITERATURE REVIEW

In this paper, the authors [6] study the smart air traffic management system and show further research while describing the shortcoming of the existing system.

Everything is achievable by technology for good health, smart living, and even smart traffic, also. IoT provides the way to improve traffic conditions and is the key player for traffic management, by using the IoT [7], the system can be enhanced which provide various aids such as smart parking, security issue, theft issue, etc. The IoT devices are embedded into the vehicle, lane intersection, signal, etc. Which provides the necessary data for smart traffic management. The IoT-based smart traffic management system provides the way to efficiently and effectively manage the traffic, this system keeps an eye over the traffic flow, traffic congestion control. By using this parameter, an IoT-based traffic management system smoothly runs the traffic management system.

The authors in [8] developed the "stochastic traffic padding" method so that opponent cannot quickly identify or attack the sensor devices which are deployed at the smart home or traffic management purpose. The developed method hides the real user activities by providing additional decoy over the user.

In [9], the authors presented the morphing technique for the traffic and this morphing technique is very useful to protect the IoT devices that are deployed in the smart environment for smartly handling the traffic since these devices are easily accessible and vulnerable to attack so attacker by using the various technologies can hack those devices And can manipulate them according to his wish.

In this paper, [10] the authors proposed the algorithm that continuously monitors the status of the conflicting lane and if algorithm finds that the lane intersection is safe, then it allows the vehicle to cross the lane, and the algorithm provides the enriched throughput, as well as it also attains the lower waiting time.

The authors [11] developed a full autonomous traffic management system for fog-based conditions, the developed system provides a fog warning and is based on green energy.

The smart city traffic management and the autonomous vehicle have been described in [12]; here, the authors proposed technology that can understand smart city traffic conditions automatically with the help of Deep Learning, AI, etc. All of this technology empowered the vehicle to be smart enough so that the necessity of the driver is eliminated. The proposed technology can also identify various kinds of traffic signals and can make the decision itself.

To effectively manage traffic conditions is not an easy task since it requires the combination of various technologies and things. The authors in [13] proposed a system that uses the traffic density as its input and this input is obtained from the camera's sensors, which are placed in various locations, and, based on the output generated, is used for signal management. Here, the system also uses an algorithm that predicts future traffic density conditions to reduce traffic, and the authors also deployed an RFID-based prioritizing system for emergency vehicles.

The authors [14] described the strategy for protected smart traffic observation, control management. Operations via embedding solution Wireless Sensor Network (WSN), Radio Frequency Identification (RFID), AdhocNetwork, and Internet of

Things (IoT) with a cryptographic algorithm. An IoT system is able to reduce computational costs and energy consumption. The authors used FOG computing to progress and deploy sensitive data inside the edge of the net and quicken feedback time for any crisis condition. The proposed system worked as a real-time scenario for detecting the parking lot, reservation, digital payment, security, etc.

In this paper, the authors [15] described IoT-based smart traffic management and a parking system is discussed.

The authors [16] described and designed the parking availability in a spontaneous manner, quickly detected the parking slot with the help of smart devices, and displayed it on the parking lot. The system reduces traffic congestion, rationalizes traffic, and enhances the parking facilities in the easiest way.

In this paper fast, improved traffic control and observing scheme proposed by the authors [17], the system capable of transferring data. The embedded system, a Vehicular Ad-hoc Network (VANET), with an execution control algorithm for the spread and maintain traffic flow via congestion avoidance, and the system includes accident precaution facility, crime detection, driver elasticity, and security of the human. The system tested by the Ns2 simulator showed an efficient outcome.

The minimum cost vehicle detects the STS system introduced by the authors [18], delivered real-time traffic update facilities. Numerous sensors integrated around the range of 500meters and via IoT castoff public traffic information and transfer as of data processed. The extensive experiments showed the effectiveness of the proposed system.

The authors [19] introduced the system named, knowledge discovery, which was able to collect data as of car-allocation positions and used, they discovered the time travel, vehicle transmission at time interval basis. The used system work on real-time scenario, so provide numerous services, i.e. analyze vehicle obtainability, instant traffic and time estimation and multimodal travel arrangements.

In this paper, the authors [20] analyzed the IoT emerged system and were able to communicate Device-to-Device (D2D). The designated system worked as automation, determined traffic density on a specific lane, ultrasonic sensors used as a data procurement component for storage of traffic compactness information and Raspberry Pi used Control Component designed for dealing out the data.

The authors [21] described the automated system MassDOT framework that given actual information during travel and spontaneously delivered to the third party designers, shielded over 700miles of state thoroughfare. The multimodal scheme integrates with the operation of the Real-Time-Traffic-Management (RTTM) arrangement.

By using autonomous vehicles, environment [22] with IOT authors tried to minimize the collision avoidance and road accident, traffic etc. the author also decrease the traffic in all toll-nakas by using the integrated system, i.e. GPS, Mobile Technologies, sensors, Wi-Fi etc. They also concentrated on fuel theft via observing the fuel and alcoholic person via detection with autonomous sensors. The authors focused on autonomous designed efficient framework able to control traffic, avoidance of car collision so decrease the accident rapidity [22].

The authors [23] described and defined the embedded STCS and numerous systems based on embedded. In the study, the authors include FES, ANN, and

WSN. They also discussed the construction, advantages, and restrictions of every approach. They worked on prevalent highly cited and test-based approaches applied to STCS. For analysis, the ANN overtook the other techniques for the learning scheme. The proposed framework worked on real-time and after studied FES very effective and WSN least costly method.

The authors [24] focused on a cognitive-traffic-management-system (CTMS) established on the IoT method, prevalent traffic lights replaced with smart traffic light incorporation arrangement. Through this arrangement, the authors optimized the traffic signal for government administration. Smart traffic light incorporation eliminates inaccuracies triggered by human influence.

The authors [25] introduced a protocol known as a Smart-Traffic-Management-Protocol (STMP), which integrates every route with numerous sensors and is accountable for observing and collecting data. Efficient observing is needed to disclose routes and city chunks jamming in a shorter interval. An exterior unit to the VANET is accountable for accomplishing and reorganizing traffic endways and paths of the city circumventing enormous traffic stages.

In this paper, the authors [26] constructed on the Discernment Based Computing architecture that enhance the surviving clarification. For this, the authors introduced an adaptive framework as of smart-traffic management. Because it was so comprehensive and complex by nature.

Traffic overcrowding, signal failures, underprivileged law enforcement, and unscrupulous traffic administration play important aspects in India and other countries; it affects the daily life and environment because most of the time management is busy solving the traffic overcrowding. Traffic overcrowding has an undesirable impression on budget, the milieu and the whole eminence of life, and most of the time management for resolving it. Numerous approaches are obtainable as traffic administration named as visual data analysis, infrared sensors, inductive loop detection, and wireless sensor networks. However, the installation of instruments and the cost of instruments is high. The authors introduced the Radio Frequency Identification (RFID) integrated with signals and leads, which decreases traffic overcrowding [27].

Innovative and automated parking and traffic system integrates with different components, i.e. procuring and data transmission. The server tackles the user request information from the database, the component manages the access and responses of the data by web services. The proposed system decreases the cost, easy access mechanism reduces the negative noises for human of the public road, the response on actual traffic, delivered the connection in between parking slots [28].

The authors [29] designed a simulative framework for a smart grid conversation scheme. The proposed framework is decentralized and automated and needed huge data and data rapidity for intelligent meeting operations; demand, supply management, and monitoring of a system are also required. Stochastic and deterministic inter-arrival periods are used for effectiveness.

A stochastic method grounded on a multidimensional Markovian prototypical, however the deterministic arrangement designated thru an innovative diagnostic method constructed on the Network Calculus. The experiments showed the effectiveness of the proposed system on GPRS, UMTS & LTE in different conditions.

The authors [30] designed a framework that comprises the investigation traffic constraints via MM1 Queuing model to optimize the service rapidity that is desirable for cultivating the long-term-evaluation (LTE) based intersection-management-system (IMS).

The authors [31] introduced traffic light environments that are able to reduce traffic jams that wedge ambulances. Transmitters and receivers play a large part in this system. Microcontroller inbuilt systems find out the position of the ambulance through GPS and transfer it to traffic light management, for wireless conversation used two Xbeepro mechanisms. The receiver accumulates for found out data, processed, computer the arbitrary position and castoff the outcome on LCD. The experiments show the efficiency of the proposed system.

The authors [32] considered the Programmable Logic Controller (PLC) technology viable in the enactment of a smart traffic light control arrangement as in the early phase. Through counting the vehicle in every frame and weight achieved the traffic-density, and also designed the automation of a parking system which was developed with specific facilities. The system emerged on highways and city traffic.

In this paper, the authors described and gave a comprehensive study of the intelligent traffic system via VANET [33]. The wireless mechanism integrates with Vehicular Ad Hoc Networks (VANETs), by which conversation of the vehicle and its infrastructure is allowed. VANET plays a crucial role in the traffic management system. The authors described VANET in the scenario of smart city scheme and concluded them the current traffic scenario designated with Intelligent Traffic Lights (ITLs) system.

The innovative traffic supervision environment abilities to decrease the congestion, propagation etc. for control the traffic route. Service degree or service interval key elements which affect traffic control. So need the Intelligent Transportation System (ITS) applications to enhance the analysis enactment. Some of the prevalent studies grounded on queuing methods and have distinctive behaviour to analyzed delay, length, inundation flow etc. according to the traffic density the service rate of the vehicle determined. The proposed system designed with traffic restrictions of an intersection thru MM1 queuing model to improve the service rapidity that is desirable to improve the long-term-evaluation (LTE) grounded intersection-management-system (IMS) [34].

The authors [35] focused on the designed visualized embedded environment through a motion sensor to detect the human who has crossed the road traffic light/ signals. If the condition meets with the system, then it stops the traffic sign and warns with an alarm to incoming motorists. The proposed system was built with high-tech motion sensors and cameras, equipped with a video image processed system. The proposed system executed a Particle Swarm Optimization (PSO) algorithm, and extensive experiments showed the effectiveness of this proposed system.

The authors described and designed [36], the 02 nodes grounded on 802.11 used for enhancement of conversational with simulates system. One-node, as mobile-node can control traffic light as worked on the path request, likewise ambulance and second-node handle roadside components. The proposed approach is able to decrease the ambulance time employed on the route as bad traffic, while offering an enhanced chance to protect patients' lives.

The authors [37] designated a traffic control system thru RFID and M-RFID mechanism. The proposed system worked on two key issues, overtaking not permitted and speed limit noticeable. The system designed with high-tech devices named as, speed-track and line sensors is able to transmit the traffic destruction to traffic management sections servers in an actual-time basis. The proposed system is under the control of the police. With these, the accidents are decreased and the law-enforcements is also followed.

In this paper, the authors [38] introduced and designed a system which is based on a real-time application for a pedestrian crossing in the city of Vienna with three automated cameras that were covered in a weatherproof housing; two cameras were used for detection and tracking, and the third one was for pedestrian-detection and tracking with stored motion movement. Arbitrary view estimation is found out via camera. The proposed work provided the guidelines for future prospects, both pedestrian and vehicle. The extensive experiments showed the effectiveness of the proposed system.

The authors [39] constructed an Instantaneous Traffic Info Arrangement grounded mobile expedients are able to find out the traffic information value and as of based on response and provide supervision to drivers. The authors described the security scenario and the need for dynamic path supervision through best elucidations. Mobile users encode the data and find out the route information and security aspect associated with s/w development time.

In this paper, the authors introduced and described the driving-traffic situations as the effectiveness of route between two ends with numerous issues [40]. Most humans used multiple kinds of online driving direction traffic arrangements, such as Yahoo maps, MapQuest, and Google Earth. The issue affected the driving traffic area route, conditions of the road, traffic on the road etc. The authors introduced a very innovative idea with the enhancement of a prevalent idea that used A* (an artificial-intelligent algorithm) and A*Traffic (a variation of A*, proposed in prevalent research). The proposed idea worked as a constraint on time in the graph depiction the route network.

The authors focused [41] on the smart traffic control system with minimum cost. There are two segments that can control configuration with an 8-kb MCS-51 microcontroller fortified by an LCD with an offset value, IC to showed calendar system and EEPROM IC stored specific time set value at numerous interval. The s/w designed to control the h/w configuration effectively, this mechanism semi-automatic. There are three frameworks used for the proposed scheme, specific time-split, time-slot period, and a specific time enhancement. The extensive experiments showed the effectiveness of the proposed system.

By novel investigation, the authors enhanced the low-cost fuel, emanations of automobiles through the custom of forthcoming data of formal traffic-lights signals, traffic flow, and deterministic traffic flow facsimiles [42]. One of the sections given in the worked grounded on forthcoming vehicle's adaptive cruise mechanism system to diminish slothful period at discontinue lights and minor use of fuel. So, the optimized control environment designed. The authors also concentrated the fuel or time-optimal rapidity curve and vehicle route assessed via a control environment. Fuel time-optimal rapidity curve designed for given the outcome of finest control delinquent.

The prevalent studies related to a topographical weight assessment scenario of a W-CDMA net costumed as new semi-smart antennas [43]. The radio-coverage tackle by used semi-smart antennas, automates system demonstrated that the dimensions enhancements, whereas the dissimilar human traffic are presented. The authors used a slight modification of the prevalent worked in a topographical restriction in an environment as multipath-propagation extant. A huge automated system designed with excellent efficiency and operative usability. The proposed scheme performed best on the topographical city atmosphere.

In this paper, the authors [44] described the safety aspects related to the traffic control scheme and discussed the operative environment, including the hardware/software framework for executing a safety precarious net because security is the main concern for controlling the operations of traffic-signal, public safety, and it is essential to safeguard for consistency of manufacturing and community approval. By custom, distributed-control mechanism enhanced the now a day's scenario via numerous set-up, i.e. automobile and pedestrian-detection and used an intelligible control mechanism.

9.6 CONCLUSION

The smart traffic system is configured by using multi-features of h/w and s/w components in IoT. Through IoT, we can enhance proficiency and optimize traffic signals. IoT devices are capable of detecting and counting vehicles on the road. Smart traffic management smartly deals with congestion, re-routing, accidents, and emergency situations.

The idea of Smart Cities is a significant vision for humankind. In recent years, great progress has been made in manufacturing smart cities in reality. The development of IoT, Big Data & Cloud technologies have given intensification to novel promises in terms of smart cities. Smart parking services and Smart traffic supervision arrangements have always been an essential part of fabricating smart cities. In this paper, we address the problem of parking spaces, emergency vehicles, crowdsourcing, and existing IoT-based Cloud cohesive smart parking arrangement. The structure after the review, we need real-time data regarding obtainability of parking slots in a parking region. Users from far-flung places could reserve a parking slot for them by the use of the smart mobile application. The determinations through in this paper are indented to expand the parking conveniences of a large city and thereby pointing to improve the eminence of life of the peoples.

REFERENCES

1. Kloess, M., & Muller, A. (2011). Simulating the Impact of Policy, Energy Prices and Technological Progress on the Passenger Car Fleet in Austria - A Model Based Analysis 2010–2050. *Energy Policy*, Volume 39, Issue 9, September 2011, Pages 5045–5062, http://www.sciencedirect.com/science/article/pii/S0301421511004629, Visited in June 2015.
2. Arnott, R., Rave, T., & Schöb, R. (2005). *Alleviating Urban Traffic Congestion*, MIT Press Books.

3. Coric, V., & Gruteser, M. (2013). Crowdsensing Maps of On-Street Parking Spaces. IEEE International Conference on Distributed Computing in Sensor Systems.
4. Jaiswal, T., Pandey, M., & Tripathi, P. (2019). IoT Empowered Smart Cities in India. 2019 Third International conference on I-SMAC (IoT in Social, Mobile, Analytics and Cloud) (I-SMAC), 249–254.
5. Jaiswal, T., Pandey, M., & Tripathi, P. (2020). Review on IoT Enabled Smart Cities in India. 2020 First International Conference on Power, Control and Computing Technologies (ICPC2T), 289–294.
6. Jiang, C. (2020). Recent Advances for Smart Air Traffic Management: An Overview.
7. Rabby, M.K., Islam, M.M., & Imon, S.M. (2019). A Review of IoT Application in a Smart Traffic Management System. 5th International Conference on Advances in Electrical Engineering (ICAEE), 280–285.
8. Apthorpe, N., Huang, D.Y., Reisman, D., Narayanan, A., & Feamster, N. (2019). Keeping the Smart Home Private with Smart(er) IoT Traffic Shaping. *Proceedings on Privacy Enhancing Technologies*, 128–148.
9. Hafeez, I., Antikainen, M., & Tarkoma, S. (2019). Protecting IoT-Environments Against Traffic Analysis Attacks with Traffic Morphing. IEEE International Conference on Pervasive Computing and Communications Workshops (PerCom Workshops), 196–201.
10. Miyim, A.M., & Muhammed, M.A. (2019). Smart Traffic Management System. 15th International Conference on Electronics, Computer and Computation (ICECCO), 1–6.
11. Majeed, F.A., Murad, B., Khalil, K., Mohamed, A.B., Ali, F., Obaid, S., & Ahmed, S. (2019). Smart Traffic Management System for Foggy Weather Conditions. 2019 Advances in Science and Engineering Technology International Conferences (ASET), 1–3.
12. Jain, S., Kumar, A., & Priyadharshini, M. (2019). Smart City: Traffic Management System Using Smart Sensor Network.
13. Javaid, S., Sufian, A., Pervaiz, S., & Tanveer, M.U. (2018). Smart Traffic Management System Using Internet of Things. 20th International Conference on Advanced Communication Technology (ICACT), 393–398.
14. Abdulkader, O.A., Bamhdi, A.M., Thayananthan, V., Jambi, K., & Alrasheedi, M. (2018). A Novel and Secure Smart Parking Management System (SPMS) Based on Integration of WSN, RFID, and IoT. 15th Learning and Technology Conference (L&T), 102–106.
15. Kaur, H., & Malhotra, J. (2018). A Review of Smart Parking System Based on Internet of Things. *International Journal of Intelligent Systems and Applications in Engineering*, 6, 248–250.
16. Goyal, R., Kumari, A., Shubham, K., & Kumar, N. (2018). IoT and XBee Based Smart Traffic Management System.
17. Rath, M. (2018). Smart Traffic Management System for Traffic Control Using Automated Mechanical and Electronic Devices.
18. Sharif, A., Li, J., Khalil, M., Kumar, R.S., Sharif, M.I., & Sharif, A. (2017). Internet of Things – Smart Traffic Management System for Smart Cities Using Big Data Analytics. 14th International Computer Conference on Wavelet Active Media Technology and Information Processing (ICCWAMTIP), 281–284.
19. Pagani, A., Bruschi, F., & Rana, V. (2017). Knowledge Discovery from Car Sharing Data for Traffic Flows Estimation. 2017 Smart City Symposium Prague (SCSP), 1–6.
20. Mahalank, S.N., Malagund, K.B., & Banakar, R.M. (2016). Device to Device Interaction Analysis in IoT Based Smart Traffic Management System: An Experimental Approach. 2016 Symposium on Colossal Data Analysis and Networking (CDAN), 1–6.
21. Bond, R., & Kanaan, A.Y. (2015). MassDOT Real Time Traffic Management System.
22. Cristina, Elena, Turcu, Vasile, Corneliu, & Octavian (2015). IoT: Smart Vehicle Management System for Effective Traffic Control and Collision Avoidance.

23. Hawi, R., & Okeyo, G. (2015). Techniques for Smart Traffic Control: An In-depth Review.

24. Miz, V., & Hahanov, V. (2014). Smart Traffic Light in Terms of the Cognitive Road Traffic Management System (CTMS) Based on the Internet of Things. Proceedings of IEEE East-West Design & Test Symposium (EWDTS 2014), 1–5.

25. Santamaria, A.F., & Sottile, C. (2014). Smart Traffic Management Protocol Based on VANET Architecture. *Advances in Electrical and Electronic Engineering*, 12, 279–288.

26. Góra, P., & Wasilewski, P. (2014). *Adaptive System for Intelligent Traffic Management in Smart Cities*. AMT.

27. Lanke, N., & Koul, S. (2013). Smart Traffic Management System. *International Journal of Computer Applications*, 75, 19–22.

28. Garcia, M.I., & García, E. (2013). Intelligent and Autonomous Management System and Associated Parking and Traffic Procedure.

29. Hägerling, C., Putzke, M., & Wietfeld, C. (2012). Traffic Engineering Analysis of Smart Grid Services in Cellular Networks. 2012 IEEE Third International Conference on Smart Grid Communications (SmartGridComm), 252–257.

30. Nafi, N.S., Hasan, M.K., & Abdallah, A.H. (2012). Traffic Flow Model for Vehicular Network. 2012 International Conference on Computer and Communication Engineering (ICCCE), 738–743.

31. Radzi, S.M. (2012). Design and Develop of Smart Traffic Light System for Emergency Vehicles.

32. Srivastava, M., Prerna, Sachin, S., Sharma, S., & Tyagi, U. (2012). Smart Traffic Control System USINGPLC and SCADA. *International Journal of Innovative Research in Science, Engineering and Technology*, 1.

33. Khekare, G.S., & Sakhare, A.V. (2012). Intelligent Traffic System for VANET: A Survey.

34. Salau A.O., & Yesufu T.K. (2020). A Probabilistic Approach to Time Allocation for Intersecting Traffic Routes. *Advances in Intelligent Systems and Computing*, vol. 1124. Springer, Singapore. https://doi.org/10.1007/978-981-15-2740-1_11

35. Matrella, G., & Marani, D. (2011). An Embedded Video Sensor for a Smart Traffic Light. 14th Euromicro Conference on Digital System Design, 769–776.

36. Mohideen, H.L., & Yyum, A.Q. (2011). Emergency Traffic System (ETS).

37. Nejati, O. (2011). Smart Recording of Traffic Violations via M-RFID. 7th International Conference on Wireless Communications, Networking and Mobile Computing, 1–4.

38. Sidla, O., Rosner, M., Ulm, M., & Schwingshackl, G. (2011). Traffic Monitoring with Distributed Smart Cameras. *Electronic Imaging*.

39. Manolopoulos, V., Tao, S., Rodríguez, S., Ismail, M., & Rusu, A. (2010). MobiTraS: A Mobile Application for a Smart Traffic System. Proceedings of the 8th IEEE International NEWCAS Conference 2010, 365–368.

40. Halaoui, H.F. (2010). Smart Traffic Online System (STOS): Presenting Road Networks with Time-Weighted Graphs. 2010 International Conference on Information Society, 349–356.

41. Purnomo, M.R., Wahab, D.A., Hassan, A., & Rahmat, R.A. (2009). Development of a Low Cost Smart Traffic Controller System.

42. Asadi, B. (2009). Predictive Energy Management in Smart Vehicles: Exploiting Traffic and Traffic Signal Preview for Fuel Saving.

43. Wang, Y., Yang, X., Lu, N., & Ma, A. (2009). *A Simulation Package for Hot-Spot Traffic Relief in WCDMA Networks*. 5th International Conference on Wireless Communications, Networking and Mobile Computing, 1–5.

44. Giri, S., & Wall, R.E. (2008). A Safety Critical Network for Distributed Smart Traffic Signals. *IEEE Instrumentation & Measurement Magazine*, 11.

10 A Robust Context and Role-Based Dynamic Access Control for Distributed Healthcare Information Systems

Abdulkadir Abdulkadir Adamu[1],
Ayodeji Olalekan Salau[2], and Li Zhiyong[1]
[1]College of Information Science and Engineering,
Hunan University Changsha, China
[2]Department of Electrical/Electronics and Computer
Engineering, Afe Babalola University, Ado-Ekiti, Nigeria

10.1 INTRODUCTION

Access control (AC) in healthcare systems is a growing concern in healthcare informatics, which has received a lot of attention over the past decade. However, not much progress has been made in this area, especially in developing countries [1]. For instance, in the Nigerian health sector and in many other Sub-Saharan African countries, little or no progress has been recorded in this area [2].

Access to healthcare information (HCI) can either be used in the positive way by authorized personnel or otherwise by unauthorized users. Controlling access to patients' and health workers' information is quite vital in the medical profession. AC for HCI systems is of importance for two main reasons: protecting patients' right to privacy and availability of the right information for medical decision-making. Availability of up-to-date patient information, coupled with inadequate infrastructure, is still an issue of concern [3]. In the advent of the availability of patients' records, access to this information needs to be controlled. A patient may easily have hundreds of separate, overlapping records in various systems. For example, at the UATH (in Nigeria's Federal Capital Territory), there exist over 100 clinical information systems that are in use. A lot of wrong diagnoses could be avoided if the systems were able to provide the relevant information required when needed. Understanding how these organizations function will help in designing better and more efficient AC mechanisms.

DOI: 10.1201/9781003140443-10

Healthcare information systems (HIS) have three possible outcomes of an access request, namely:

    i. Normal access granted
    ii. Emergency access granted (normal access denied)
    iii. Access denied

This work aims to integrate contextual information and dynamic properties into Role-Based Access Control (RBAC) to give a better AC model for HIS, thereby minimizing the need for AC exceptions.

The remaining sections of this paper are organized as follows. Section 2 presents the related works. The research methodology is presented in Section 3, while the system design and architecture are presented in Section 4, and in Section 5, the results are discussed. Finally, in Section 6 we conclude the chapter.

10.2 RELATED WORKS

AC models are categorized into two major classes, namely: those based on AC lists (ACLs) and those based on capabilities. In a capability-based model, access is granted based on capability; access is conveyed to another party by transmitting such a capability over a secure channel.

In the ACL-based model, a user's access to information depends on whether the user's identity is on a list of authorized users. AC models are sometimes categorized as either discretionary or non-discretionary. The three most commonly used models are Discretionary Access Control (DAC), Mandatory Access Control (MAC), and Role-Based Access Control (RBAC). MAC and RBAC are both non-discretionary [4].

10.2.1 ROLE-BASED ACCESS CONTROL

RBAC is usually not determined by the user but by the system. RBAC is useful in military operations, commercial software, and applications where multi-level security requirements are a necessity [5]. The difference between RBAC and DAC is that the latter allows users to control their resources, while RBAC does not [6]. RBAC can be distinguished from MAC primarily in the way permissions are handled. MAC controls read and write permissions based on a user's clearance level and additional labels. RBAC controls collections of permissions that may include complex operations, such as an e-commerce transaction, or may be as simple as reading or writing. A role in RBAC can be viewed as a set of permissions [7], [8].

Three primary rules are defined for RBAC, namely:

1. Role assignment: A user can execute a transaction only if the user has been selected or has been assigned a role.
2. Role authorization: A user's active role must be authorized for the user. With rule 1 above, this rule ensures that users can take on only roles for which they are authorized.

3. Transaction authorization: A user can execute a transaction only if the transaction is authorized as the user's active role. With rules 1 and 2, this rule ensures that users can execute only transactions for which they are authorized.

Additional constraints may be applied as well, and roles can be combined in a hierarchy where higher-level roles subsume permissions owned by sub-roles. Most IT vendors offer RBAC in one or more products.

10.2.1.1 State-of-the-Art in RBAC Model

Despite all of the advancements in research with respect to AC systems, one clear fact still stands out; the RBAC system still remains a very dynamic AC system upon which many other AC models are built [9], [10]. The concept of RBAC was first formally proposed by David F. Ferraiolo and D. Richard Kuhn in their paper [11]. Up until then, AC had largely been based on one of two principles: mandatory access control (MAC), enforced by the system, or discretionary access control (DAC), where users are allowed to determine who should get what kind of access to the objects they have created.

The inventors of RBAC found that DAC and MAC had some shortcomings, mainly that:

- DAC is flexible for the users but makes AC hard to administrate.
- MAC on the other hand gives the administrators total control and no flexibility for the users.
- Neither of them scales very well as an organization grows in the number of employees and/or computerized resources.

RBAC was proposed in an attempt to solve these problems. The key concept in RBAC is to define roles that correspond to job titles in an organization. Each role is associated with a set of access rights. Employees holding the same job within an organization are assigned to the same role – thus, the number of roles is considerably lower than the number of employees in an organization.

The key benefits of RBAC are:

- Ease of administration – the AC reflects the structure and responsibilities within an organization and the number of roles is smaller than the number of employees.
- Scalable – more roles can be added as new job titles and responsibilities are added.
- Flexible – if the responsibility of an existing job title changes, it is only necessary to change the access rights of the corresponding role, not for every single employee holding this job title.

RBAC has gained increasing popularity over the last decade. It has proven to be especially well-suited for organizations that are highly dynamic: where people

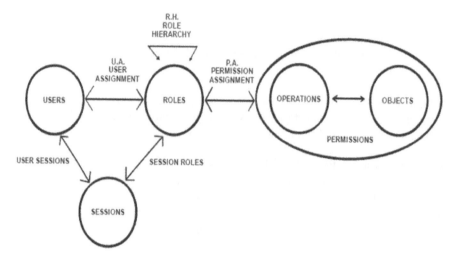

FIGURE 10.1 Role-based access control (RBAC) model.

come and go and change responsibilities often and where there are vast amounts of information. A typical RBAC model is shown in Figure 10.1.

10.2.1.2 Benefits of RBAC

The use of RBAC provides many advantages to organizations. Hierarchical support is used by roles to easily reflect the organization's structure [12]. RBAC supports the access permission delegation through roles. Separation of duty avoids role conflict among users and restricts the users in acquiring conflicting roles at the same time. For example, a doctor can either acquire a day-doctor or night-doctor role but not both at the same time.

10.2.2 OTHER ACCESS CONTROL MODELS

Sodiya and Onashoga [13] discussed the following AC models:

- Purpose-Based Access Control (PBAC): In this case, access is granted based on the intentions of the users. Each user is required to state his or her access purpose when trying to access an object. For example, in a school environment, data is collected for registration, checking of results, and so on. The system validates the stated access purpose by the user to make sure that the user is indeed allowed for the access purpose.
- History-based Access Control (HBAC): This describes an AC technique in which access is granted based on the previous records. A user is granted access to an object if the user has had previous access to the object to some reasonable threshold.
- Temporal Constraints Access Control (TCAC): This involves AC policies in which time restrictions were attached to resource access. For example, some activities must be performed within a reasonable time frame.

However, most of the current AC models are not completely adequate to ensure effective AC to computer resources because they are still faced with some problems. Some of these are:

- Difficulty in tailoring access based on various attributes or constraints;
- Difficulty in encapsulating all possible job functions and requirements to access objects;
- Inadequate capability of the administrator to compose all rules that cover the necessary access constraints and permission between users, operations, and objects because of dynamic nature of operation;
- Non-prevention of unauthorized access;
- Denial of authorized access because of complicated rules [14].

10.2.3   ROLE HIERARCHY

Role hierarchy is the structural representation of roles that reflects the organizational work breakdown structure. Some roles are classified as super role and sub-role. The hierarchical relation among sub-roles is represented by the super roles. Typical examples of sub-roles of a doctor include the roles of cardiologists, neurologists, and medical specialists. The main purpose of role hierarchy is to allow administrators to appropriately state the AC policies and to assign them to the super roles [15].

10.2.3.1   Separation of Duty (SoD)

Separation of Duty (SoD) plays an important role in an AC policy specification. Role hierarchy allows permission inheritance that sometimes allows users to acquire conflicting permissions at the same time. SoD is a way of resolving these conflicting issues so that a user "S" should not acquire conflicting permissions "P" because of the role hierarchy. For example, a doctor cannot be his own patient.

10.3   RESEARCH METHODOLOGY

In this section, we present an overview of how this research work was carried out. This study involved four different phases. The first phase was studying the health information system of the UATH, while the second phase was building a computerized database for the health records of the hospital. The third stage was creating and infusing a database AC system based on context, using a DRBAC. The fourth phase was to network the servers from the clinical departments (where patients' health records are generated) into an intranet such that access can be gained both from within the hospital by authorized officials only after they have been authenticated.

10.3.1   CONCEPTUAL DESIGN

This study is experimental research, wherein the research team not only observes but manipulates the studied subjects. The subjects here refer to data i.e. health records of patients in the hospital. In this context, the term "manipulation" does not mean tampering with or changing the actual record already filed for a patient; rather it means transforming data from hard to soft copies which can be better administered with access to very sensitive information properly controlled while ensuring that only authorized persons get approval and under the right circumstances get access to such information regarding patients health records.

10.3.1.1   Population of the Study

The data collected for this study is made up of health records of patients in the UATH and data of hospital personnel who use the hospital's computing environment. The hospital is located in the Gwagwalada Area Council of Abuja (Nigeria's Federal Capital Territory).

10.3.1.2   Sample Size of the Study

A sample size of about 5,000 patient records was collected. The hospital caters to a wide range of specialist services at subsidized rates by the federal government of Nigeria. Out of the 5,000 patient records, we have concentrated on 63 records (3 each from the 21 clinical departments of the hospital) to create a patient database, while a sample of 30 members of staff of the hospital was drawn from a total of 400 to create a system user database.

10.3.1.3   Sampling Technique

The technique used is the stratified sampling technique. This technique is a modification of the well-known random sampling technique, wherein every element in the population has an equal probability of being selected. Using stratified sampling, the entire population under study (herein referred to as health records and hospital staff) is divided into various departments from where they originated. Simple random samples are then drawn from each department.

10.3.2   Data Collection Instruments

The data used for this study was obtained from a combination of two methods: face-to-face interviews with three principal officers of the hospital, namely: the head of administration, head of personnel, and the medical records unit, as well as a personal observation, and secondly, assessment of data storage facilities in the records unit of each of the 21 clinical departments.

10.3.3   Validation of Data Collection Instruments

The validation procedure used was based on physical validation. The instrument used was an interview, which was subject to scrutiny by the project manager before being administered. This involved comparing the items (i.e. the questions) on the interview form with the objectives of the study to ascertain whether a correlation existed. The interview items were scrutinized to ensure that the data provided by the respondents would answer the research questions needed for the study. After the items were verified, the instrument was validated and the study team was allowed to proceed with the study.

10.3.4 Data Collection Procedure

The interviews were scheduled for three separate days with the heads of administration, personnel, and medical records, respectively. The interviews took place in their offices in the afternoons when they were really free to answer all the questions required of them by the research team. The head of personnel provided the data required to create the user database, while the head of medical records took the study team around the registries of the 21 clinical departments where patients' records were randomly selected and used to build the patient database upon which the research work was done.

10.3.5 Procedure for Data Analysis

Data collected from the personnel and medical records unit was presented in a tabular form and was used to build two databases (system users and patients) using the Microsoft Access 2007 software and a Graphical User Interface (GUI), which is used to access the database using the Microsoft Visual Studio 2008 software. The authentication process to allow system users to access the health records from the database was based on the context, and the purpose for which authorized users seek information of a patient's health record.

10.3.6 Setting Up the UATH Computing Environment

The UATH has very few computing devices to support its activities. There are a few host-based applications; a few systems are client-server based and are accessible over the LAN using PCs. From the 400 permanent employees, student nurses, contractual staff, and 150 trainee doctors, a username and password were created for a sample of 30 hospital staff in the UATH domain for the purpose of accessing computing resources.

A domain keeps a record of objects that have been grouped together for the purpose of management [16]. The direct or indirect objects of the object list are differentiated by a domain. To build up and maintain a structure for the UATH network, standard servers such as DNS, DHCP, and FTP were put in place. The IT department is responsible for the management of all IT resources.

10.3.7 Systems Functional Requirement Analysis

An appropriate RBAC model based on purpose and context was developed while suitable hardware and software platforms for the study were acquired and designed, respectively. The following steps were taken to accomplish this:

1. Mapping of the AC system to fit organizational structure by studying the available documentation regarding the operation of the systems in the UATH computing environment.
2. Formulation of a set of requirements after studying the UATH computing environment, the organizational structure of the hospital, and the way its

activities are carried out with reference to the separation of duties and delegation of authority.

10.3.7.1 Grouping of Users

Users are grouped based on the RBAC96 model to stress the principle of least privileges and separation of duties [16]. These privileges give rise to three entities, namely: users (U), roles (R), and permissions (P). The objective of grouping users is to provide a basis to form an AC method that closely models how UATH operates and allows each user to accomplish his or her assigned task.

10.3.7.2 User Definition

A user can be a human being, a host, or a process. In this study, the term "User" refers to people who interface with the computer system. These people include resident doctors, nurses, consultants, trainee doctors, admin, IT staff, and any other persons authorized to use the UATH computing resources.

10.3.7.3 Role Definition

A role is a job function or job specification within an organization, which defines the authority and responsibility conferred on individuals performing that role. Simply speaking, the roles define the responsibilities and authorities individuals have. In this study, the roles were defined based on the responsibility of the user. The role was defined by function and not by the position of the individuals within the organization. Every user is assigned at least one of the roles depending on his or her job functions.

10.3.7.4 Permission Definition

Permission is described as the right to access or an approval to access information on a system. Permissions can apply to single objects or to many objects. For example, permission can be specific to give a user access to a particular file or to all files. Each role is assigned one or more permissions. Users gain permissions through their respective role assignments. Permissions can be revoked from roles when conditions warrant it.

10.3.7.5 Resource Definition

Categorization of resources was done in terms of hardware, software applications, and data (files and folders). Roles were given access to resources by assigning them appropriate permissions.

10.3.8 Analysis of Existing RBAC Models

This section gives the limitation of the existing models such as ARBAC and ERBAC, which seem to have all the requirements outlined in Section 3.7. However, the limitations of these models when applied to the UATH environment are given in this section. These existing models are based on NIST RBAC. Users are given permissions using their assigned roles as shown in Figure 10.2.

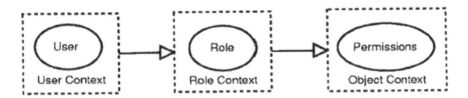

FIGURE 10.2   The core entities of RBAC.

The core of the available RBAC models consists of three entities: users, roles, and permissions. The model assumes that relation r1 will change more rapidly than relation r2. But if r2 changes fast the operation will not go as structured. Administrators will have to spend a lot of time administering the r2 relation. However, there are two problems with this arrangement [17].

First, each time permission is changed, removed, or added, all affected permission-to-role assignments need to be changed. In a system where permissions change often, this can lead to plenty of extra work for the administrators and, thus, the objectives of the study will not be met. Second, if the number of permissions is high, administrators may not be familiar with all permissions and this may lead to mistakes when permission-to-role assignments are made.

The UATH environment at present does not have an adequate number of systems and resources. Considering the various existing RBAC models and the kind of network structure we proposed, it was discovered that the existing models do not support the system of permission assignment used in those systems. Therefore, it was necessary to come up with a new model for AC to address these issues.

10.3.9   MODIFICATION OF EXISTING RBAC MODELS

The major problem with existing models as indicated in the previous section is the relationship between roles and permissions. This work is primarily concentrated in this relationship and the result is a new abstraction level between roles and permissions. The newly developed model makes an attempt to add this by introducing the notion of context and purpose in accessing resources from the various organizational units. Resources are targets upon which AC is applied. These include data files and hardware as well as office and business application software.

Each user is assigned one or more roles, and each role is assigned some permissions. The organizational units reflect the structure of the hospital management. Every employee within the organization has rights to access resources, but the purpose for such and the context (i.e. the situation) under which such is requested must first be ascertained before permission is granted for access. Thus, it is not only the role assigned to a user that matters. Purpose and context come into play in the proposed model using some domain-specific constraints incorporated into the RBAC model. The proposed model is shown in Figure 10.3.

Users are arranged on a flat structure with no dependencies upon each other. When a user logs on to the UATH network, a role is selected. This is the role the user has been assigned to. With the role come the permissions which the user can

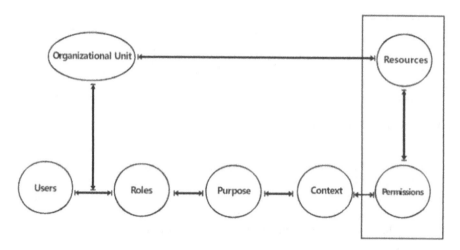

FIGURE 10.3   Proposed RBAC model with the incorporation of purpose and context.

exercise on resources, most especially data and software applications. Several roles may be assigned to a single user in an organization. Users can be assigned to several organizational units.

10.3.10   SYSTEM ARCHITECTURE AND THE TECHNICAL APPROACH TO SYSTEM DESIGN

After fully aligning the service delivery requirements of UATH with the specific objectives of the study, a modular approach to the architectural design of the system was employed based on layers. This was done to ensure the functionality of modules in a given layer without affecting the performance of other layers. A system model was developed for this purpose. A detailed description of all the components involved in the design and implementation of the model is presented in Section 4.

10.3.10.1   Software Design Platform

The .NET framework was the platform used to implement the system after a comprehensive analysis of its capabilities in relation to other frameworks like J2EE as an alternative for the implementation. The .NET platform is preferred to other platforms mainly because of its in-built capability to effect data synchronization as opposed to building it from scratch.

10.4   SYSTEM DESIGN AND ARCHITECTURE

10.4.1   THE BASIC RBAC SYSTEM STRUCTURE

The tool designed is an RBACS. Going by the UATH organizational policy, roles are defined as a combination of the official position and job function. Typical

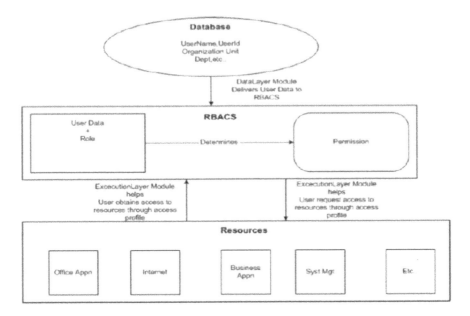

FIGURE 10.4 The proposed RBACS structure.

official positions could be that of unit heads, consultants, specialists or resident doctors, nurses, trainees, pharmacists, laboratory technologists, admin, and IT staff. The job function is simply what things they do by holding that position.

All the data are defined and maintained in a human resource database system. Figure 10.4 shows the proposed architecture of the system and its interfaces to the database and individual applications.

The database stores user information e.g. username, password, user-level, role, and so on. The user information is what the system uses to determine which permissions a user has to access resources.

Ideally, each employee is only assigned to one role. However, in special circumstances, an employee can be given more than one role, for example when they serve in acting positions or in case of illness of a colleague. In that case, the acting employee will be assigned more than one role or by assigning the acting employee a higher role in the role hierarchy. In this study, the hierarchical assignment of roles was used and, thus, several applications can be accessed through one role.

10.4.2 Technical Approach to System Design

The system was developed based on the layered model architecture. This model as discussed is sometimes called the abstract machine model. It organizes the system into layers, each of which provides a set of services. The RBACS was developed using four layers, which are discussed below and also shown in Figure 10.5.

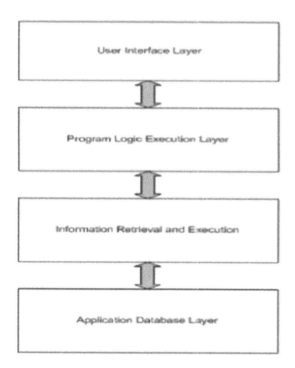

FIGURE 10.5   The layered model of the RBAC system.

1. **User Interface**
   This is a graphical user interface that allows the user to interact with the system. It allows the requester of the resource to issue queries and request resources to be allocated.
2. **Program logic execution layer**
   This constitutes all the business logic of the application. It constitutes the user authentication and authorization modules plus the resource allocation and management modules.
3. **Information Retrieval and Execution (Data Processing Layer)**
   This layer constitutes all the application logic that handles all the interactions between the application and the security database.
4. **Application Database**
   The application database used is a Microsoft Office Access 2007 database that stores and retrieves security credentials. New users, roles, and permissions are added to the database.

10.4.3   CODE ARCHITECTURE

10.4.3.1   Role-Based Security in .NET

This study's application was designed using the .NET code architecture. On this architecture, role-based security can be easily implemented. Security enforcement comprises two main stages, authentication and authorization. Authentication is verifying the user's identity [18]. The application verifies that a user is a person they claim to be. The authentication method adopted for this study is for users to enter their username (of between five and ten characters) and passwords (made up of letters and digits). The application looks up the user record in the database with the username that a user entered, and then verifies that they entered the matching password. Without successful authentication, users are not allowed to enter the application.

This study's application provides access to the user through identity authentication through a principal. Identity and Principal were used to limit what the user can do within the application. Identity identifies the user by username while Principal combines Identity with a list of roles that the user plays. Identity and Principal enables one to divide areas of responsibility by assigning different users to different roles, and they allow users to share areas of responsibility where needed by assigning more than one user to the same role.

10.4.3.2 The Authentication and Authorization Process Using Identity and Principal Objects

The Identity and Principal objects are used in determining what permissions a user has on a system. An Identity is an object that contains a user name. At the fundamental stages, identity objects contain a name and an authentication type. The name can either be a user's name or a Windows account name, while the authentication type can be either a supported logon protocol, such as Kerberos V5, or a custom value. The Identity object typically represents the logged-on user, but it can represent any user whose rights are used to determine what tasks are allowed to be performed. Microsoft.NET provides a set of standard Identity objects representing users logged on to the system from different logon services; WindowsIdentity, FormsIdentity, PassportIdentity, or GenericIdentity. All Identity objects have in common a useful property called Name, which returns the name of the currently logged-on user.

10.4.4 System Implementation

10.4.4.1 System Implementation Environment

RBAC models are typically environment-independent. For example, RBAC can be implemented at the application level or embedded in the operating system. RBACS implementations discussed in this work were demonstrated in the UATH's network environment. Some of the objects which are controlled by RBACS were spread among several servers connected to the network.

AC mechanisms require that security attributes be kept about users and about objects. For a user to access any resource, be it on the network or the desktop computer, he will have to go through this tool and he will be granted access to do only what his job entails and nothing more.

To make this project tool fully functional, windows group policies were set up. One was to turn off windows X hotkeys and all shortcuts to command prompt to prevent users from accessing resources using keyboard commands like "RUN." Second, a domain/computer user logon group policy was necessary to enable this project tool to run automatically at user logon so that users have no other means of accessing computing resources other than through this tool.

10.4.4.2   Main Features of the Developed Software

The software is a menu-driven computer-based program. It is a very robust and powerful data handling system for the management of the hospital. The use of the software is, however, not restricted to this design as it can also be modified. It is a user-friendly package that is composed of many forms, folders/files.

The various containing forms are discussed as follows:

frmAbout: This form gives brief information about the software and the computer.
frmAccounts: This form allows the account personnel to input the cost incurred by the patient to the database
frmChangepassword: This form allows the user to change his password.
frmlaboratory: This form allows the laboratory personnel to input the laboratory tests done by the patient and its cost into the database.
frmlogon: This form allows users to logon to the software.
frmpatientreg: This form allows the admin personnel to register the patient to the database.
frmPharmacy: This form allows the pharmacists to input the drugs prescribe.
frmtreatment: This form allows the medical personnel to diagnose the patient and give him another date for next appointment is needed.
frmMain: This form is the main form where the menus are selected.

The developed software interface is shown in Figure 10.6.

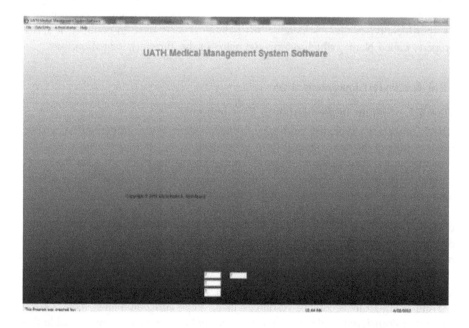

FIGURE 10.6   Interface of developed software.

10.5   RESULTS AND DISCUSSION

10.5.1   System Testing

The system was demonstrated and tested using the collected data.

10.5.2   Database

The database stores system user's information, associated roles, and permissions. Any database management system would suit this project, but Microsoft Office Access 2007 was used to develop the database because of its simplicity. Figure 10.7 shows the developed patient database. The information which can be found in the database includes: Card Number, Name of Patient, Occupation, Contact Address, Residential Address, Contact Number, Date of Birth, Sex, Blood Group, Geno-Type, Marital Status, and passport photograph of the patient.

10.5.3   Relationships Between the System User Database Tables

The system user database comprises five tables as shown in Figure 10.8. The User table stores system user information; Security Rights stores permissions and Security Groups stores roles that have been allocated to system users.

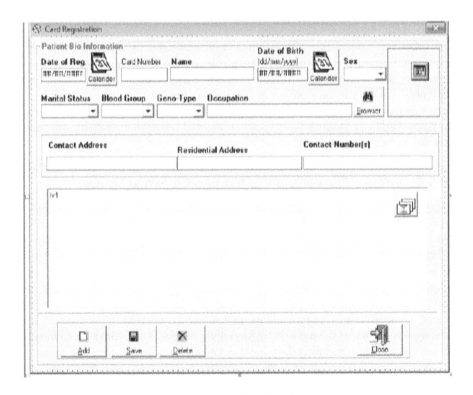

FIGURE 10.7   Screenshot of the patient database interface.

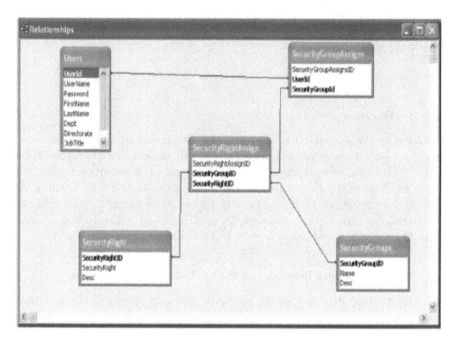

FIGURE 10.8   System user database tables used for the RBAC model.

The other two tables, Security Right Assign and Security Group Assign are, respectively, used to store permission to role assignments and role to user assignments. The authorization process checks the username and its password and compares it with the information in the User table in the database. After successful authentication, it reads the user information from the User table and from the Security Group Assign table to ascertain which role (i.e. security group) a user has been assigned.

10.5.4   Infusing Purpose and Context into the System Design

As mentioned earlier, documents, files, and folders created by a user can only be saved in the server allocated to that department, hence a user cannot access files and folders from other departments. It should be noted that "users" in this sense refers to doctors and nurses in the various departments of the hospital. Also, all-access modes discussed so far are used for normal purposes. However, considering the nature of the medical profession, there will sometimes be a need for doctors in one department to access the information of a patient, who has been treated or been undergoing treatment in another department. This kind of information is known as the medical history of the patient. This information enables the doctors currently handling the patient to know the various stages of an illness a patient has passed through and the various forms of treatment administered to such a patient in the past. This will help the doctor provide better treatment to the patient. Should the case of a patient become critical, such that it necessitates the referral of the case to a specialist hospital (in the country or overseas) for more intensive care, the doctor who makes such reference will need to make use of the patient's medical history to write their report.

Given the present setting, only the head of the unit currently handling a patient's case is granted access to information from other departments. However, there will be times when the head of the unit is absent or indisposed and the urgency of the situation will require the doctor handling the case to access information to which they do not have normal access. Clearly understanding that denying access to such information may place the patient's life in great danger, the concept of an administrator to provide "some users" with access to information which they cannot access normally was added to the AC system. These special privileges are, however, given only to "senior doctors."

Thus, there are two purposes for which a doctor (who has only been assigned the user role) might require information from other departments. These are reference and research purposes. The situations (i.e. the context) under which a doctor might need to make reference to the medical history of a patient are as follows: when trying to improve treatment; during an emergency; when referring the patient for better treatment elsewhere; or when conducting an investigation into the medical processes that led to the loss of a vital organ, incapacitation, extreme adverse reaction of the patient to a drug or, in the worst case, the death of a patient.

Granting access to a doctor attempting to access patients' medical history for the purpose of research, will only be possible under these situations: if there is a serious case which requires an in-depth study [on how such cases were successfully handled in the past] before deciding on which course of action to take; if the doctor is a staff of UATH and has been required by the hospital management to carry out a research geared toward improving the treatment of certain diseases; if the doctor (though not a staff of UATH) is an expert in a particular field in medical practice and a member of a professional body in the medical field who has been sent by an external body (such as a government agency, a health society or an international organization) to conduct a research for or if the doctor (being a staff of UATH) has been granted permission by the hospital management to carry out a personal research (using the hospitals' resources) to help advance their academic qualifications and medical practice.

It must be noted that for both purposes; reference and research, approval must have been given by the hospital management to the benefitting doctors through their unit heads. This is done after they have signed an agreement to use the information accessed only for the approved purposes and not to violate in any way, the privacy rights of the patients whose medical records will be accessed. The system administrators must also be informed of such approvals to enable them to make adequate preparations. After such approval is granted, the system administrators create a user account in the database for all doctors with these special privileges, and a special privilege password is issued to them.

The mode of operation of this purpose-context-RBAC is simple. Logging in a "user" will not provide them with the permissions enabling them to view everything except what is allowed through enabling the menu items.

10.5.5 ANALYSIS OF HYPOTHESES

The major aim of this research is to devise a means by which AC in HIS can be significantly improved. In an attempt to realize this aim, we came up with a hypothesis, hoping possibly to verify and was thus put forth:

AC in HIS can be significantly improved by developing and implementing an AC model that extends the core RBAC principle by including contextual information and dynamic properties in AC decisions.

Though broad and rather generalized, this hypothesis was further divided into sub-hypotheses to properly verify them. They were as follows:

H1: Existing AC models used in healthcare systems can be significantly improved by making use of information relevant to AC, like context of use and dynamic properties.
H2: Involving end users in the requirements capturing process for AC will provide information that may increase the knowledge to be utilized in designing a better AC model.
H3: There exists contextual information in a healthcare setting that can be included in an AC model.
H4: Contextual information can be ranked according to relevance and usefulness for AC.
H5: There exist dynamic properties related to use of information systems in healthcare that can be included in an AC model.
H6: Dynamic properties can be ranked according to relevance and usefulness for AC.

Based on the AC model we have developed, we can analyze each of these sub-hypotheses one after the other starting with the first.

H1: Existing AC models used in healthcare systems can be significantly improved by making use of information relevant to AC, like the context of use and dynamic properties.

As has been demonstrated in the model presented in this research, it is sometimes necessary to press for further information about system users before access to data is granted. This inclusion of context is useful in determining whether the user (even if authorized) should be granted access based on the circumstances under which he seeks permission. Dynamic properties also signify that changes can be made at every point in time and this can also affect AC decisions. For instance, if a system-user who previously was a trainee doctor and had limited access to patients' medical records completes their training and attains the full status of a resident doctor, there has to be a way of taking care of this change in status such that they are now able to access more information. Also, in the case of a doctor leaving the hospital, a subordinate can be promoted to the office vacated by the outgoing doctor, and with this new change comes the privilege to access more information while yet ensuring that patients' information privacy rights are not violated.

H2: Involving end users in the requirements capturing process for AC will provide information that may increase the knowledge to be utilized in designing a better AC model.

Of course, after the AC system has been designed, tested, and implemented, it is the system users that will be at the receiving end. They will be the ones to detect any loopholes in the system since they are the ones who constantly find themselves in different situations (pleasant and unpleasant) when trying to access data for which they should have access (or already have perhaps). As such, they will be able to advise the software developers in such a way that they will consider all the problems users might face in trying to utilize the AC system since one thing these developers seek to achieve is user-friendliness. Thus, as we have done in this research, understudying the UATH computing environment, consulting a number of unit heads and personnel in charge of patients' medical records provided us with valuable information, which enabled us to identify possible shortfalls and to make arrangements to cover for them.

H3: There exists contextual information in a healthcare setting that can be included in an AC model.

As can be seen from the last sub-section under the system testing section, there actually exists contextual information that can be included in an AC model in a healthcare setting. Basically, especially as was improvised by the research team here, inclusions were made of purpose and context. Two purposes were identified – reference and research.

Various contexts were formulated based on the information supplied by the head of the medical records unit and some other personnel in charge of patients' records. A time constraint was also involved, thus putting some control over the period of the working day during which system users can access information under various circumstances. It is only for emergency purposes that authorized users can information on a 24/7 basis. It should be noted in this research work that all the five user roles assigned to system users allow for normal access and the context is seen as being "a part of their routine work schedules" while some users with special privileges are granted alternative access wherein they have to specify their purpose and context before access is granted.

H4: Contextual information can be ranked according to relevance and usefulness for AC.

Looking at the section that describes the alternative login process, there was a clear ranking of contextual information. The purpose was first declared before context was specified. Under the context listing, a ranking was done such that any specification above is of more importance than the one following under. There were some other contexts identified but were left out in the selection process because they are not relevant to AC in a hospital setting or they rarely occur.

H5: There exist dynamic properties related to the use of information systems in healthcare that can be included in an AC model.

These dynamic properties are the changes that occur in the organizational structure of the healthcare institution, especially as we have it in a distributed healthcare system operated by UATH. As time goes on, some personnel rise up the ladder giving way for subordinates to fill in the offices they have left vacant. Some leave to other institutions and, thus, create a vacancy that must be filled. Again, the leadership of the organization might be replaced by a new board which will introduce some new changes to the mode of operation of the hospital.

Sometimes, even when no one is leaving or being promoted to a higher position, the organizational structure might just be changed for some reason or the other, or perhaps, new clinical units (involving new operations, new technologies, and more staff) are introduced which might create challenges to the database AC system in operation. The question then becomes: how can we retain a good AC mechanism considering the changes that have been made and might still be made at any point in time?

In view of this, roles were created to make up for these changes which may come up as time progresses. As changes are made, hospital personnel were simply assigned new roles which guaranteed them increased access rights to the information stored in the patient database. Therefore, whether or not sudden or frequent changes are made, patients' medical records remain safe and secure such that a new change will not allow an employee excessive access to patient information while denying some who should have been granted access.

H6: Dynamic properties can be ranked according to relevance and usefulness for AC.

As earlier said, roles were created to make up for these changes (dynamic properties) that usually occur in healthcare settings.

10.5.6 Testing of Hypothesis

Having analyzed each of the six sub-hypotheses coined out of the major hypothesis, we were able to verify and accept each of them based on proofs from the study conducted.

With the aim of devising a means by which AC in HIS can be significantly improved, we came up with the null and alternate hypotheses given as follows:

H0: AC in HIS can be significantly improved by developing and implementing an AC model that extends the core RBAC-principle by including contextual information and dynamic properties in AC decisions.

H1: AC in HIS cannot be significantly improved by developing and implementing an AC model that extends the core RBAC-principle by including contextual information and dynamic properties in AC decisions.

Based on our findings, we, therefore, accept the null hypothesis and reject the alternate hypothesis.

10.6 CONCLUSION

This chapter presented a new dynamic RBAC method to effectively grant access to authorized hospital personnel. This research motivation came from the complicated access control (AC) requirements inherent in current healthcare data management systems. The findings of this study will be of great help to healthcare organizations, businesses, corporate organizations, financial institutions, and government establishments which make use of networked computer systems to automate their work processes and allocate roles to employees.

REFERENCES

1. De Carvalho Junior MA, Bandiera-paiva P (2018) Health information system role-based acess control current security trends and challenges. *Journal of Healthcare Engineering*, pp. 1–8. DOI: 10.1155/2018/6510249
2. Odekunle, FF, Odekunle RO, and Shankar S (2017) Why sub-Saharan Africa lags in electronic health record adoption and possible strategies to increase its adoption in this region. *Int J Health Sci (Qassim)*, vol. 11, no. 4, pp. 59–64.
3. Treasurer, LL (2015) Infrastructure – the key to healthcare improvement. *Future Hosp J*. vol. 2, no. 1, pp. 4–7. DOI: 10.7861/futurehosp.2-1-4
4. Bakker AB (2004) Access to EHR and access control at a moment in the past: a discussion of the need and an exploration of the consequences. *International Journal of Medical Informatics*, vol. 73, pp. 267–270.
5. Ferraiolo DF, Sandhu R, Gavrila S, Kuhn R, Chandramouli R (2001) Proposed NIST standard for role-based access control. *ACM Transactions on Information and System Security*, vol. 4, no. 3, pp. 224–274.
6. Shin MS, Jeon HS, Ju YW, Lee BJ, Jeong S (2015) Constructing RBAC based security model in u-healthcare service platform. *Scientific World Journal*, pp. 1–13. DOI:10.1155/2015/937914
7. Yarmand MH, Sartipi K, Down DG (2013) Behavior-based access control for distributed healthcare systems. *Journal of Computer Security*, vol. 21, no. 1, pp. 1–39. DOI: 10.3233/jcs-2012-0454
8. Zhang L, Ahn GJ, Chu BT (2002) A role-based delegation framework for healthcare information systems. Proceedings of the Seventh ACM Symposium on Access Control Models and Technologies - SACMAT '02. DOI: 10.1145/507711.507731
9. Yeo SS, Kim SJ, Cho DE (2014) Dynamic access control model for security client services in smart grid. *International Journal of Distributed Sensor Networks*, vol. 10, no. 6, pp. 181760. DOI: 10.1155/2014/181760
10. Ferreira A *et al.* (2007) Access control: How can it improve patients' healthcare? *Study in Health Technology and Informatics*, vol. 3, pp. 65–76.
11. Ferraiolo DF, Kuhn DR (1992) Role-based access controls. ArXiv, abs/0903.2171.
12. Tahir MN (2009) *C-RBAC: Contextual role-based access control model.* Multimedia University, Malaysia, pp. 62–75.
13. Sodiya AS, Onashoga AS (2009) *Components-based access control architecture. Issues in Informing Science and Information Technology*, Department of Computer Science, University of Agriculture, Abeokuta, Nigeria, pp. 700–704.
14. Kayes ASM, Kalaria R, Sarker IH, Islam MS, Watters PA, Ng A, … Kumara I (2020) A survey of context-aware access control mechanisms for cloud and fog networks: taxonomy and open research issues. *Sensors*, vol. 20, no. 9, p. 2464. DOI: 10.3390/s2 0092464
15. Hu J, Weaver AC (2005) *A Security Infrastructure for distributed healthcare applications.* Department of Computer Science, University of Virginia, Charlottesville, VA, pp. 20–31.
16. Hornos MJ, Rodríguez-Domínguez C (2018) Increasing user confidence in intelligent environments. *Journal of Reliable Intelligent Environment*, vol. 4, pp. 71–73. DOI: 10.1007/s40860-018-0063-4
17. Georgiadis CK, Mavridis I, Nikolakopoulou G, Pangalos GI (2002) Implementing context and team based access control in healthcare intranets. *Medical Informatics and the Internet in Medicine*, vol. 27, no. 3, pp. 185–201.
18. Dhillon PK, Kalra S (2018) Multi-factor user authentication scheme for IoT-based healthcare services. *Journal of Reliable Intelligent Environment*, vol. 4, pp. 141–160. DOI: 10.1007/s40860-018-0062-5

11 Impact of ICT on Handicrafts Marketing in Delhi NCR Region

Archita Nandi

Department of Research Innovation & Consultancy,
I. K. Gujral Punjab Technical University, Kapurthala

11.1 INTRODUCTION

India is a country with immense cultural heritage, and the past of Indian crafts has been prosperous for many decades. "Past testifies to the fact that India's artisans were globally famous for their expertise and craftsmanship". The carvings on the temples offer evidence of the expertise, experience, and participation of artisan craftsmen in their art. Since time immemorial, exports of Indian handmade products have taken place. This industry, which forms a significant part of the country's rich cultural heritage, uses the traditional skills of artisans in various crafts such as woodenware, metalware, textile weaving and printing, marble and stone crafts, leather works, jewelry, etc. This expertise is passed down in the context of family practice from generation to generation. True to its name, instead of modern technologies for producing diverse products, the "Handicraft" (handmade crafts) field utilizes traditional manual methods. It is an unorganized, decentralized cottage industry that is labor-intensive. The cultural richness of India offers a lot of remarkable art and craft items. The Indian craft industry is labor-intensive and is typically decentralized and distributed across the community. While the handicraft industry hires millions of artisans, with regard to the global industry, it is still minuscule. With the introduction of several websites supporting the Indian craft industry's e-commerce, this situation is evolving. The government has also launched numerous efforts to provide market openings for artisans doing business in the unorganized field [1–5].

Handicraft represents the culture and skills of the local people and, hence, of the country. Due to differences in culture and individuals who manufacture different kinds of crafts, India is one of the most sought-after destinations for crafts. Different places in India are popular for various handicrafts, such as Jaipuri quilts in Rajasthan, embroidered fabric in Gujarat, Phulkari in Punjab, wrought iron items in Jodhpur, woolen products in Kashmir, etc. One of the main employers in rural India is the craft sector [6–10]. Some artisans operate with their hands on a full-time basis

DOI: 10.1201/9781003140443-11

153

and many on a part-time basis to manufacture these items. Any of the factors that are allowing this sector to expand faster are low initial expenditure, export opportunities, and international earnings. The vision for the Handicrafts Sector for the Government of India's 12th Fifth Year Plan is to establish a globally successful handicrafts sector and provide artisans with sustainable livelihood opportunities by creative product designs, product quality enhancements, the adoption of new technologies, and the preservation of traditions. To grow and show the hidden skill and handicraft of India, IoT devices create an enormous amount of data allowing vendors to gather more data points about their users through websites, blogs, and YouTube channels. The availability of data will let marketers test out innovative ideas which further and help grow their business. With the use of IoT devices, investors determine hidden skills, customer requirements, and behavior traits of their users. All of this can be used to enhance the marketing efforts further Internet of Things (IoT) as "the interconnectivity of our digital devices that provides endless opportunities for creators to listen and respond to the requirements of their customers – with the accurate message, at the correct time, on the exact device." The digitalization of data plays a very important role here, in which data mining and data analytics are used to find out the right market for customers as well as for the creators.

Industry analysts agree that coupled with conventional methods of exchange, world trade now relies mostly on e-commerce. While there is no specific concept of e-commerce, e-commerce transactions are characterized by the OECD (Organization for Economic Cooperation and Development) as the selling or purchasing of products or services, whether by computer-mediated networks, between companies, households, individuals, governments, and other public or private organizations. Goods and services are ordered via such networks, but the billing and actual distribution of the good or service can be carried out on or off-line.

According to the UNESCO-UNCTAD/ITC, "Artisanal goods are the concept of handicrafts created by artisans either entirely by hand tools or also by mechanical means, as long as the artisans' direct manual contribution remains the most important." The distinctive characteristics of artisanal products originate from their distinctive characteristics, which may be practical, aesthetic, imaginative, culturally connected, decorative, practical, conventional, symbolic, and essentially religious and social. The government's emphasis in the early 1970s was on the protection of crafts, capacity enhancement, and education, resolving the artisans' livelihood issues. The government changed its emphasis in the 1980s and 1990s to improve and tap the export market's capacity. The goal was to enhance the socioeconomic conditions of artisans by creating opportunities for jobs. Metalware, woodware, hand-printed textiles, embroidered and crocheted items, shawls, carpets, bamboo products, Zari products, fake jewelry, sculptures, earthenware, jute products, marble sculpture, bronze sculpture, leather products, and other handicrafts are part of the Indian Handicraft Industry. It is possible to clarify the sector's demand in terms of the market size (Figures 11.1 and 11.2).

For an economy like India, handicraft businesses are the most ideal: low capital usage, strong job density, and broad regional dispersion. It is an organization that

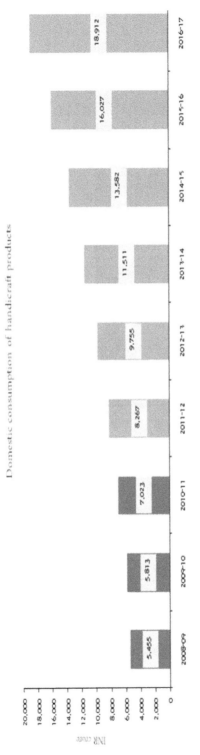

FIGURE 11.1 Domestic consumption of handicraft products in India [7].

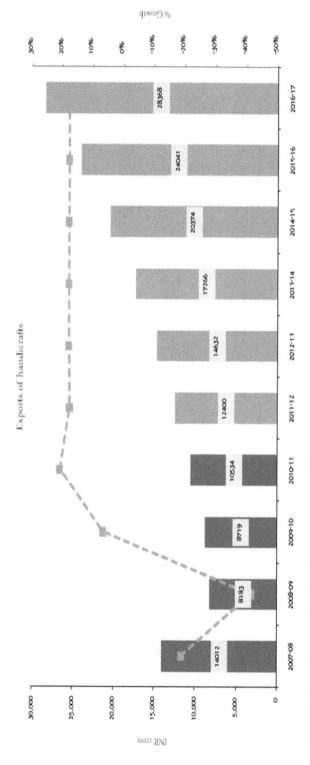

FIGURE 11.2 Exports of handicraft products in India [8].

can produce tremendous jobs with limited usage of resources, but the business is troubled by several issues. Finance, marketing, and innovations are the three major factors. There is still very little technical interference in the selling of the handicraft industry's products. The key restrictions are as mentioned below in the handicraft field.

- Lack of design, innovation, and technology upgradation
- Highly fragmented industry, unstructured and individualized production systems
- Limited capitalization, access to credit, and low investment
- Insufficient market information on export trends, opportunities, and prices
- Limited resources for production, distribution, and marketing
- Limited e-commerce competence among producer groups

Lack of adequate infrastructure, absence of latest technology.

11.2 INFORMATION AND COMMUNICATION TECHNOLOGIES (ICT) IN HANDICRAFT

India is one of the world's biggest Internet consumers and is anticipated to move over to the US in the coming years. Therefore, all companies and local artisans have a tremendous secret and untapped potential for e-commerce. The way business is conducted and transacted by offering a global marketplace has shifted. E-commerce offers a perfect distribution outlet for goods manufactured domestically. It also offers exporters space for growth. E-commerce operations fill the far broader distance between artisanal clusters in India and the future consumer.

Information and Communication Technologies (ICT) will enable artisans and handicraft/handloom organizations to transcend regional barriers and link to a broader client base. An increasingly digital world that involves 980 million smartphone subscribers and 350 million Internet users is evident in India's ICT trends. By 2020, India's e-commerce industry is projected to hit USD 100 billion (Live Mint, 2015). India's online businesses endorse the Government of India's "Made in India" campaign, while enabling small local manufacturers to bind to a broader consumer base, including artisans. Digital technology will get them exposure, consumer knowledge, potential clients, and enable them to sell more, even digitally, to businesses and organizations based in rural areas and with less capital.

The term ICT applies to the usage of a broad variety of networking devices and computerized information systems. Items and services such as personal computers, notebooks, mobile devices, wired or wireless intranet & Internet are included in these technologies. ICT uses corporate productivity tools such as text editor and spreadsheets, business software, data and security storage, network security, and others. The handicraft industry demand would have more prospects and possibilities than ever before with e-commerce and e-business.

The craft industry is a highly innovative field that provides a broad range of artisanal goods. Yet rural producers/suppliers are currently failing, for numerous reasons, to sell their goods and services. Because of their illiteracy, ignorance, and insecurity, artisans dependent on intermediaries for raw materials, financing, and the demand for finished goods. In accordance with the tastes and desires of buyers, the popularity of handicrafts depends on the goods launched on the market. To build artistic and ethnic appeal with a touch of creativity, industrialization of all forms of arts and crafts must be accepted by the customer industry by the exchange of goods. Some challenges are addressed by artists and these are licensing concerns, low demand, lack of access to raw materials, obsolete machinery and equipment, lack of qualified labor, high excise, weak facilities, policy instability, shipping problems with local transport (up to port), low incentives, income tax, Internet accessibility, promotion, among these big marketing issues. Rural producers/suppliers are currently failing, for numerous reasons, to sell their goods and services. Marketing implies communicating with consumers rather than any other company activity, of which the core of modern marketing is to create consumer connexions focused on customer desire and loyalty (Armstrong and Kotler, 2010). Marketing is characterized as the method of producing, delivering, promoting, and selling products, services, and ideas in a competitive atmosphere to promote satisfying trade ties with consumers (Pride and Ferrell, 2003), and there is no knowledge available to artisans regarding the use of information technology for marketing. One other big explanation for Laggard's usage of information technology by artisans is that India has a poor share of the world's craft industry. In the case of India, the artisans are not conscious of sophisticated or frontier marketing techniques and they have no knowledge to purchase and sell about Internet use. India's artisanal rivals such as China, Thailand, Korea, etc., on the other side, have new and superior handicraft and marketing technologies relative to India's handicraft and marketing activities. The craftsmen of Haryana faced the same dilemma, too. In the promotion of any commodity, information technology has played an important part. Information technology supports operations, along with their associated methods, management, and implementation, involving the development, storing, handling, and communication of information (Adeleke, 1985). The pace at which information technology is being built would transform the marketing analysis method for handicraft goods entirely. The 2004 program, however, stressed the sense of technology from the point of view of daily life, culture, business, and the environment, as well as technology dependence. An active dialogue on the position of technology education and information technology has taken place during the last 20 years. Nevertheless, the advent of information technology would bring about exponential progress and success in marketing research. Data is the lifeblood of both an individual and an organization. The development and longevity of any company are pivotal and important. In marketing, knowledge is both a stimulus and a medium by which the goods, services, and concepts of companies have achieved the intended destination (Onuoha, 1998). Communication technology in several respects impacts marketing. Some of these save labor and provide facilities. Others develop radically different goods and new ways of organization. Others are also improving the usage of the Internet as the strongest way to meet target consumers. The future benefit of IT to handicrafts can be seen by cost savings, performance enhancement, and improvement in productivity. In the first place, it is important to evaluate and record the knowledge needs of artisans, and then to establish appropriate information systems (IS). The emphasis of systems implementation is on current issues resulting from the modernization and globalization of the handicraft industry. To improve the perceptions and actions of buyers, it is about exchanging knowledge between buyer and seller. Media technology entails advertisement, promotion of purchases, personal sales, and advertising of handicraft items.

As India opens its doors to multinationals during the age of economic change and liberalized markets, putting an end to the Raj license, it is not just the economies that always encounter in the domain of the global economy, but also the citizens and cultures that add a new dimension to the multicultural world. The wonder of globalization is that there is still a cross-cultural exchange between the "local" and the "national" throughout modern times, and amid several inconsistencies, the global village is now not just an opportunity, but a fact. The boundaries between world cultures are now, needless to say, eroding and becoming meaningless. The whole planet is linked today. Nevertheless, globalization today is still branded as a push toward profit-making. The theories of cultural/economic globalization define it as the western substitution of local culture. "When Coca-Cola or the chicken of Kentucky seek the right to open branches in China or Russia, their primary purpose is not affection for the citizens of Russia or China, but the benefits they would gain from such an undertaking." What is significant is not the availability of McDonald's and Kentucky's products in New Delhi or Mumbai in the days of globalization and for the citizens of India, but rather the availability and accessibility of South Indian Dosha, North Indian Rajma, and Orissan Dalma in the streets of New York, Stockholm, and Toronto. Again, it is often of the belief that the global-local interface gives plurality/hybridity to cultural trends rather than substitutes. But the accommodation toward assimilation is there.

On the other side, speaking of the Indian Handicrafts landscape, it is time for a complete assistance, both financial and skill-based, from the government to the local yet valuable craftsmen of our great cultural heritage, as nothing can substitute it once lost. Strong government help is also required for illiterate poor artists and vulnerable crafts, not half-heartedly with a removal intent, but with earnest and real efforts. In this industry, thus, the government should invest more resources supporting and selling the crafts and supplying artists with instruction. Moreover, numerous cooperatives and charitable organizations ought to make sincere attempts to strengthen the artists' working conditions. DWARAKA (Development of Weavers and Rural Artisans in Kalamkari Art) e.g. is an organization that helps hundreds of Andhra Pradesh Kalamkari artists by providing loans for their children's education and providing medical and marriage expenses. Similarly, through its numerous exhibits, "Dastakar" and NGO allow professional artisans from different states to market their products.

One of the biggest problems is that the handicraft base is very unorganized in the Indian handicraft industry. Modern instruments and tools with which the handicraft base is very weak are used by the artisans. Therefore, to increase the standard of production, the expertise of artists that should be provided with sufficient raw materials and appropriate financial assistance should be strengthened. Innovation should be taken care of at the same time, but it should not weed out originality. Besides, special attention is required for the promotion and export of the goods. The craftsmen should also be adequately introduced to the business, leaving the intermediaries with little space for interference. The Ministry of Media, Trade, and Tourism should jointly take measures to render craft goods globally known and commercially viable.

And last but not least, care should be taken to popularize craft goods that build recognition among home customers in home markets. This would boost demand as well as supply. Handicraft goods should be produced with the new trend and style of customers in mind, so that ethnic crafts with contemporary characteristics can be readily adopted. However, to preserve the originality of the goods, consideration should be paid. It is significant to notice that international organizations such as the World Bank, the ADB, and UNESCO, etc., refer to the craft and cultural industries as an increasingly significant source of work growth and the production of sustainable profits, and that funding for cultural industries should be redefined as development investment rather than expenditure.

Originally began as a part-time hobby among rural people, crafts have now developed into a booming industry with substantial and sustainable consumer demand and significant economic potential. Needless to mention, handicrafts are currently making a significant contribution to work generation and exports from the region. It has, however, played an essential role in India's economy and culture. There are ample prospects for Indian hand-crafted goods made by their professional artists in both the domestic and international markets with the advent of globalization. But risk-prone geography still remains. To save and promote this unique cultural asset of the vast country, a concerted effort by citizens, government, and civil society is, therefore, required.

In the Delhi NCR area, the handicraft industry has a significant presence with products such as art metalware, imitation jewelry, wood products, leather footwear, terracotta, etc. This would render the NCR region a potential location for the domestic and global market for the development, exchange, and export of handicrafts. The significance of the NCR area in terms of handicrafts is further increased by a large number of tourist inflows.

11.3   USES OF INFORMATION TECHNOLOGY FOR MARKETING

11.3.1   Advertising of Handicrafts products

First of all, in ads, artists should utilize information technologies. The companies have been involved in advertisement, exhibits, publishing of brochures, and participating in trade fairs for the marketing of crafts. The business often utilizes the Internet as an advertisement platform, such as Facebook, WhatsApp, and other social networking tools, such as pop-up ads, banner ads, and search engine ads. Promotion of handicraft products: Promotion of sales requires a broad variety of strategies to include an added short-term motivation or incentive to patronize a shop or purchase a commodity. Techniques include samples, premiums, vouchers, games and tournaments, exhibits and box inserts, demos, and shows. Sales promotion is usually meant to boost the other variables in the promotion mix and to maximize their short-term efficacy. With the aid of the expertise of online musicians, the usage of social media will provide consumers with exclusive offers for sales.

E-tailing is still in its developing process in India and provides e-tailers with a massive opportunity due to growing disposable incomes, more Internet and smartphone penetration, and increasing customer perception of e-tailing. It is also a positive thing for art vendors as it provides a new opportunity for them with consumers' growing handicraft tastes. This e-tailing potential needs to be utilized by handicraft organizations as it gives them a modern route to expand their business presence. They will need to reach semi-rural and rural areas to be effective. Initiatives can be taken by the government and e-tailing firms to render the supply chain and payment processes problem-free. To meet more consumers, mobile applications should be designed. Handicraft manufacturers who do not have much knowledge of e-commerce businesses can link up with e-commerce businesses such as Flipkart, SnapDeal, Amazon, eBay, etc.

11.4   TOOLS OF INFORMATION TECHNOLOGY

As marketers today, we are blessed to have a wide range of free and low-cost resources that allow us to learn about our consumers, rivals, and markets. These online resources also assist us to effectively operate our business.

- Autosend: You will instantly deliver customized, tailored email and text messages to a user depending on what they do on your website if you wish to chat face-to-face with each entity that enters your website. So you will proactively give a personal note to them if a consumer arrives, makes a transaction, or simply looks at a relevant FAQ query on your website.
- PromoRepublic: It is a tool for free posting. It delivers more than 100,000 ideas, models, and graphics to consumers for amazing content on Facebook, Instagram, Twitter, and LinkedIn. A team of experienced programmers and copywriters makes all the social network models. With a built-in graphics editor, users can quickly configure models. Drag-and-drop graphics software to modify beautiful models and build content from scratch. Auto-posting and scheduling to Facebook, Instagram, Twitter, LinkedIn. With the Promo Republic, you get a library of 100,000 post templates and visuals and a calendar of post ideas for any day of holidays, days from past, hot topics, and activities.
- Chattypeople: In the marketing environment this year, the chatbox has made its impact. In the past decade, chatbox technology has notably advanced, empowering advertisers to build boxes without any coding expertise. For anyone looking to build a Messenger chatbox easily, it is the perfect method. The app makes it easy for developers to build a box that fits

Facebook, pushing consumer deals on demand. You would be able to use Facebook Messenger and feedback to drive clients into the sales funnel with chat boxes. Take orders from Facebook directly, gather details on the clients and keep them up-to-date with specific items,

- Hello Bar: Hello Bar is a basic tool that lets you turn guests into consumers by quickly designing banners that showcase your customers with your most relevant material, goods, services, and notifications. Everything you've got to do is log in and pick a target.
- Typeform: It is a technique that needs to be tried while searching for innovative forms to engage with future clients.

11.5   CHALLENGES FOR E-TAILING IN HANDICRAFTS MARKETING

Most of the handicraft e-tailing projects were not as profitable as they were supposed to be, the primary factors were:

- Safety threats – Security issues are at the forefront of discussion as it comes to customer worries when shopping in the online media for handmade goods. Many customers are discouraged from making online transactions through a lack of confidence and privacy issues. Consumers are often worried about the usage during Internet purchases of their personal details supplied to them.
- Customer retention – Several customers shopping on the Internet do it out of interest and this renders it extremely difficult to shop repeatedly. If the consumer is not happy with his first impression of purchasing artisan goods online, he will not make a repeat order.
- In the case of product categories needing comparatively greater consumer engagement, the e-tailing route is found to be vastly ineffective in supplying customers with necessary details. The retailing of items such as homemade garments, natural cosmetics, etc. are examples. Since the details needed for making these transactions and consumer participation is minimal, most consumers are happy purchasing books and music on the Internet. In the case of a blue trouser, though, the client likes to know stuff like: What hue of blue is it? What does the skin feel over it? How fast is it going to crease? Such a dilemma would not concern conventional retailing. The Internet provides small quantities of essential knowledge to the user in the non-standard commodity categories. In such instances, the true condition of the trouser is only revealed to the vendor and this contributes to "knowledge asymmetry."
- Shopping is indeed an activity of touch-feel-hear – others do not suffer from "time-poverty" and shopping is still treated as a family outing. Therefore, this sort of atmosphere produces a consumer retention challenge.
- Complicated medium – The simplicity of use is a concern, as in certain instances the site design can suffer from high difficulty bordering on utter

chaos. The haphazard architecture of online web portals for handicrafts is not user-friendly and does not result in sales.

- Compared to physical art shops and model catalogues, e-tail stores have no uniform designs. For each craft e-tail shop, thus, numerous user habits (navigation schemes) need to be taught. This is a transient concern as the growth of the Internet progresses.
- Flaws of website design – For craft online platforms, visual appearance, and aesthetics may not be as convincing as in the case of a physical department shop or a product catalogue. This is a transient dilemma that can be solved by the evolution of web design.
- Restricted Internet service – Not all consumers, as they do in the mail system, have access to the Internet. This is a transient concern as the growth of the Internet progresses.

11.6 CONCLUSION

The handicraft sector is highly labor-intensive, cottage-based, and decentralized. The sector is spread throughout the nation, mainly in rural and urban areas. Most of the handicraft units are located in rural and small towns and in all Indian cities and abroad, there is huge commercial potential. For rural communities, the handicraft industry is a major source of income, employing more than six million artisans, including a large number of women and people belonging to the weaker segments of society. However, the lack of accessibility to artisans of communications facilities such as computers, Internet, and telecommunication systems, etc., is a major handicraft market problem. Second, the level of literacy of artisans is quite low and the behavior of consumers is sort of traditional, which is a problem for effective communication. The study concluded that craft products have enormous opportunities and the need to reach or communicate with artisans about the use of information technology tools to boost marketing is felt.

REFERENCES

1. Arijit Ghosh, "Initiatives in ICT for Rural Development: An Indian Perspective", *Global Media Journal – Indian Edition*, Vol. 2 (2), 2011, pp: 11–19.
2. Atanu Garai, B. Shadrach, "Taking ICT to Every Indian Village: Opportunities and Challenges", One World South Asia, 2006, pp: xiii.
3. Jamshed Siddiqui, "A Framework For ICT Adoption In Indian SMES: Issues And Challenges", *International Journal of Information Technology & Management Information System*, Vol. 4 (3), 2013, pp: 114–120.
4. Shreya Jadhav, "Indian Handicrafts: Growing or Depleting?", *IOSR Journal of Business and Management*, Vol. 2 (15), 2013, pp: 07–13.
5. Suhail M. Ghouse, "Indian Handicraft Industry: Problems And Strategies", *International Journal of Management Research and Review*, Vol. 2 (7), 2012, pp: 1183–1199.
6. Suhail Mohammad Ghouse, "Export Competitiveness of India: The Role of MSMEs to Play!", *International Journal of Management Research & Review*, Vol. 4 (11), 2014, pp: 1069–1084.

7. S.V. Akilandeeswari, C. Pitchai, "Study of Handicraft Marketing Strategies of Artisans in Uttar Pradesh and Its Implications", *Research Journal of Management Sciences*, Vol. 2 (2), 2013, pp: 23–26.

8. S. Venkataramanaiah, N. Ganesh Kumar, "Building Competitiveness: A Case of Handicrafts Manufacturing Cluster Units", *Indore Management Journal*, Vol 3 (2), 2011.

9. Towseef Mohi-ud-din, Lateef Ahmad Mir, Sangram Bhushan, "An Analysis of Current Scenario and Contribution of Handicrafts in Indian Economy", *Journal of Economics and Sustainable Development*, Vol. 5 (9), 2014.

10. Waqar Ahmad Khan, Zeeshan Amir, "Study of Handicraft Marketing Strategies of Artisans in Uttar Pradesh and Its Implications", *Research Journal of Management Sciences*, Vol. 2 (2), 2013, pp: 23–26.

12 Intelligent Amalgamation of Blockchain Technology with Industry 4.0 to Improve Security

Sumit Kumar Rana and Sanjeev Kumar Rana
Department of Computer Science and Engineering
Maharishi Markandeshwar (Deemed to be University)

12.1 INTRODUCTION

We live in a technology era, as we saw the evolution of different emerging technologies. These technologies include the Internet of Things, Artificial Intelligence, Machine Learning, Cloud technology, etc. The technologies have an impact on our conventional approach of process execution in multiple dimensions. Integration of these technologies with current infrastructure results in new possibilities for the development of new business model. This integration resulted in an industrial revolution, which is called the fourth industrial revolution, which includes smart infrastructure like smart machines, smart logistics, smart cyber-physical systems, etc. Smart infrastructure means all the components of the infrastructure are connected and these components can communicate or exchange information with each other without human intervention. This smart infrastructure offered various advantages like automation of the workflow, high efficiency, quality, and error-free execution. This industrial revolution will initiate the operative effectiveness and will also cultivate the possibilities to raise the client value.

Although this fourth industrial revolution can provide multiple benefits, there are some issues that must be tackled for better utilization of the smart infrastructure. These issues are because of the exchange of information between different components. Some of the issues are security and privacy of data, trust among communicating parties, traceability of data, etc. Thus, new technology was required to deal with these issues.

During this fourth industrial revolution, one more emerging technology was taking a fly named blockchain technology. This technology came into consideration after the huge success of a cryptocurrency named Bitcoin. Bitcoin was built on the

DOI: 10.1201/9781003140443-12

concept of blockchain technology. In a blockchain, there are groups of peers that represent different organizations of parties involved in the communication or exchange of information. All the transactions that involve the exchange of information are stored in a decentralized and immutable ledger. Decentralized means no single peer can control this ledger and immutable means once any transaction is written in the ledger it cannot be removed. Because of this immutability attribute, peers can trace the transaction back to its origin. Another feature that makes it more popular is consensus. Transactions will be entered in the ledger after the consensus of the communicating parties so each peer will be aware of the information exchange. After exploring this technology, researchers suggested that blockchain technology has the potential to uplift industry 4.0 to a new level in various domains. Integration of this technology with industry 4.0 will open up new development opportunities. The abovementioned attributes of blockchain technology can help in optimizing the existing industrial processes. Using blockchain technology, new business models can be formulated that are completely interconnected and secure. Next, we will discuss the history of revolution in the industrial domain that took us to the fourth generation of the industrial revolution.

12.2   HISTORICAL LOOK BACK OF INDUSTRIAL REVOLUTION

Modern business has seen a rise since its earliest iteration at the start of the economic revolution within the eighteenth century. For hundreds of years, most of the production of clothes, food, etc. was done with hand tools and with the help of animals. By the end of the eighteenth century, this manual system was replaced by a manufacturing system. From here onwards the industries move on to reach up to their current state. Let us have a look at this evolution from 1.0 to 4.0 in brief.

12.2.1   INDUSTRY 1.0

Industry 1.0 started in the eighteenth century, when mechanical production is done with the employment of steam power. Before that, it was like spinning wheels were used manually for the preparation of threads. Employment of steam power gave wings to the first industrial revolution (Figure 12.1). Steam-driven machines

Mechanical Production Tools
        and Equipment

FIGURE 12.1   Industry 1.0.

increased the production volume by tenfold so businesses were able to serve the demand of a large customer base. The main achievement of that time was the steam engine, developed by James Watt in 1776.

12.2.2 INDUSTRY 2.0

Industry 2.0 started in the nineteenth century with the invention of electrical energy also known as electricity. Then that electrical energy was used to run machines. Those electrical machines were better in performance with a less cost as compared to the steam-driven machines. Before electrical machines, production was done on a one-by-one basis which was very slow and inefficient. The concept of the assembly line was introduced during this time span (see Figure 12.2). Those electrically driven assembly lines were capable of mass production. This model of production transformed the process in almost all industrial domains. Various approaches were proposed which used this assembly line model to improve the efficiency of the overall process and lower the cost of production.

12.2.3 INDUSTRY 3.0

Electronic industries saw an abrupt change after the development of integrated circuits and logical controllers. That advancement in the electronic industry gave wings to the third industrial revolution. Integration of these electronic devices brought automation to the production process (see Figure 12.3). Automation brought accuracy, speed, and, also, reduced the effort required for the production process.

Along with the electronic devices, software solutions were also employed, which enabled proper management and tracking of the resources and process. These software solutions are continuously updated to incorporate functionalities as per need.

Electrical Facilities and Assembly
Lines

FIGURE 12.2   Industry 2.0.

Automated Production with
electronic devices

FIGURE 12.3   Industry 3.0.

Connectivity with smart technologies

FIGURE 12.4   Industry 4.0.

12.2.4   INDUSTRY 4.0

Exponential change in the Internet and its connected technologies resulted in the fourth industrial revolution. Machines that were equipped with electronic devices are supported with networking technology. Now, machines can communicate and exchange information without the intervention of human beings. A new concept of the smart factory came into existence, which includes smart machines, smart logistics, and more. These smart factories utilize the latest technologies like the Internet of Things, Artificial Intelligence, etc., for a complete digitized production environment (see Figure 12.4). If we take the example of IoT only, this technology has a huge impact on the rise of the fourth industrial revolution. The main objective of employing IoT in consideration is the connectivity among industrial things like machines, tools, etc. This connectivity helps in the automation of the industrial process. After going through this journey of the industrial revolution, we can definitely say that industry 4.0 is here for a long stay. Next, we will discuss the work done by different researchers in associated fields of our chapter as our literature review.

12.3   LITERATURE REVIEW

In this [1], the authors discussed the recent research directions in the context of the fourth industrial revolution and how blockchain can be used to implement the same. They also elaborated on the difficulties that can be faced while implementing it with the help of blockchain. New domains explored to find the applications of this emerging technology. Further, they believed that their research would provide the launchpad to empower the new researchers and industry professionals.

In this [2], the authors explained the possibilities of various privacy attacks that can take place in different core components of industry 4.0 infrastructure. They discussed various issues related to privacy, information stealing, and the importance of awareness about confidentiality. At last, the effect of breaches in personal data in this data-oriented era is discussed.

In this [3] article, researchers developed a new approach for communication between different parties. Multiple channels are created to pass and receive information with better security. They adopted blockchain technology for the exchange of information so this new approach for communication is considered consistent and safe as compared to the traditional approach.

In this [4] paper, the authors elaborated the various research dimensions to improve cybersecurity by incorporating the emerging blockchain technology. As online systems can be penetrated easily in today's technologically advanced environment, an approach is required that uses cryptography for better security arrangements. This is why blockchain is in high demand for organizations to support their cyber-physical infrastructure.

In this [5] paper, the authors analyzed issues of the automotive industry faced by stakeholders with the conventional approaches. Then they examined the applicability of the blockchain in this field to uplift the automotive industries. After discussing positive, negative, and various issues, the authors provided some suggestions for the researchers to further investigate the same domain.

In this [6] paper, the authors reviewed an application for industry 4.0, which was supported by blockchain technology. Then they provided the description about the advantages and the issues occurred during the development of blockchain supported application for industry 4.0. This study can be used by young researchers to go in deep to enhance the security in the industry 4.0 domain.

The authors proposed [7] a credit-based consensus algorithm to ensure the safety of an IoT device-supported system. An access control method was designed by the authors to safeguard the data collected from the IoT device sensors. This system used a directed acyclic graph with blockchain technology. They implemented the proposed model with Raspberry Pi.

In this [8] paper, researchers elaborated the applicability of blockchain technology in the manufacturing domain to make it smarter. They proposed a model supported by blockchain to ensure safer, consistent, and self-directed manufacturing. This will bring trust among the various stakeholders of the manufacturing system.

The authors [9] consider security and privacy as the main areas to work on to make a smart system more robust. This smart system is prone to various cyber-attacks. The authors reviewed blockchain technology as a solution that can be incorporated in the industry 4.0 environment to protect against cyber-attacks.

In this [10] paper, the author proposed a blockchain-supported Cyber-Physical System and proved that their system is capable of providing better security to protect against different cyber-attacks. Then after implementing the proposed model they performed an analysis of its performance.

Initially, the author [11] analyzed the conventional models of blockchain for various applications. Further, they explored the areas related to industry 4.0 where blockchain technology can be integrated for the betterment. Then they discussed the limitations of conventional security approaches in comparison with the blockchain-supported solutions.

In this [12] article, the author described how the blockchain and other emerging technologies like AI, IoT, etc. are being integrated to devise smart solutions for the fourth industrial revolution. These smart solutions are changing complicated and time-consuming processes into fast and easy tasks.

In this [13] article, the author discussed the complete environment of block-chain technology. They presented a comparative analysis of all the trade-offs of this emerging field. Then they explained various components of blockchain technology like consensus algorithm, smart contracts, etc., and future research directions in this domain.

Authors in [14] presented a study of various domains where blockchain can support the automation of multiple tasks. They explored various research articles from scientific journals of high repute. They found many research gaps that can be explored by researchers to carry out studies in the domain of blockchain technology.

In this [15] article, the authors identified different advantages and limitations of integrating blockchain technology in industrial domains. Further, they investigated various tools and platforms that can be used for implementation for the same. Still few issues are undiscovered that can support better utilization of the technology.

In this [16] article, the authors proposed a decentralized solution for the storage of information received from IoT networks. Also, they discussed a consensus algorithm for the circulation of resources among various nodes of IoT networks. Then, they presented the comparative analysis of the proposed approach with available approaches.

In this [17] paper, the authors discussed the security and privacy parameter in the context of blockchain. Firstly, the authors presented the concepts of blockchain and its applications in different domains. Finally, they demonstrated how blockchain technology can be incorporated within various fields to provide security and privacy by discussing the case of the Bitcoin cryptocurrency.

This article [18] provided an understanding of security services being used in current applications in various domains. The authors presented the issues faced with current security solutions and discussed how blockchain technology can be used to overcome these issues. Further, different blockchain-supported applications are compared on the basis of security services delivered by these applications.

In this [19] article, the authors explored the use of blockchain in the development of decentralized access control for the collaborative environment where organizations cannot trust each other. They also investigated the same in supply chain use cases.

12.4   SECURITY PREFERENCES FOR INDUSTRY 4.0

Industry 4.0 can be thought of as a framework where smart tools and technologies are integrated to provide smart solutions, smart products, smart supply chain, etc., as shown in Figure 12.5.

There are few principles that an industry 4.0 solution should follow but these principles have some flaws, which are discussed below:

1. Communication: All the tools, machines, etc. being used in the industry framework must have the capability of communication so that these machines can exchange information as per need without human intervention. Information will be exchanged with the help of some networks, like the

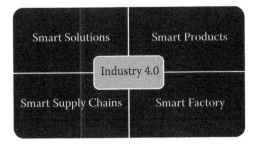

FIGURE 12.5   Industry 4.0 framework.

Internet, IoT [20], etc. Information sent over the network has the possibility of eavesdropping or modification so this will bring security issues.

2. Service-Oriented: Industry 4.0 tools and machines should be accessible on-line for different services as per need. These access requests can be within the organization or from remote locations. IoT [21] will play an important role in this context. If any service can be operated from a remote location then the possibility of a denial-of-service attack is, thereby, generating multiple fake access requests.

3. Decentralization: Generally, a centralized approach is used where a central authority will take all the decisions. In the case of a collaborative environment, multiple organizations can work together on some projects. No organization can trust each other blindly, so a trust issue will arise. Thus, a decentralized approach is required.

4. Real-time Capability: Machines and tools should be able to gather, analyze the information, and take necessary action. If it would require some third party to resolve the issue then it will cause a delay in the process. It should be capable of detecting faults or any issues in the process without the involvement of any third party. It should have some smart facility that can be triggered by an event.

5. Scalable: It should have the scalability option to expand or change the existing system process as per the requirement of the organizations.

6. Provenance: Sometimes a small change or fault can be the cause of the big issues of the system. If any of the issues arise in the process then there must be some facility so that history logs of the process can be traced back to its origin and find the root cause of the issue.

Let us take an example from the automotive industry to understand the principles mentioned above. In the automotive industry, multiple companies collaborate for production e.g. a car. This car has multiple parts from multiple manufacturers. As with the fourth industrial revolution, the production process has been changed completely. Most of the process is automated now and human intervention is very less because of the implementation of IoT [22–24]. Process, which is automatic and operating using some networks, is more prone to cyber-attacks. Cyber-attacks can cause production downtime, spoilage of end products, financial loss, etc. Due to an abrupt growth in the number of cybersecurity cases cyber resilient technology is the primary concern. So, a solution is required to make this process more secure and attack-proof.

As per the cost of a data breach report 2019 of IBM [24], a data breach caused by any cyber attack can bring a huge loss to the organization. The top four sectors that suffered from the highest loss due to data breach is health, financial, energy, and industrial sector. If we consider the industrial sector only, the total cost of a data breach was found to be around USD 5.20 million, which is a huge amount. With proper security measures, this can be reduced to a significant extent.

12.5   HOW BLOCKCHAIN CATERS SECURITY TO SUPPORT INDUSTRY 4.0

Blockchain is one of the emerging technologies that will have an impact on most of the industrial domains. It is basically a chain of blocks that are connected using the hashing technique. These blocks store the transactions performed over the peer-to-peer network. Information once written in the block is immutable and visible to all the concerned nodes of the network. All the transactions need a consensus of all the concerned nodes before execution.

Blockchain technology can help in various ways like automation of a process, integration of various entities, tracking and tracing of processes, cryptographically secure information exchange, etc. However, exploration of the benefits of this technology in the industrial domain is still in the inception phase. Still, we can find some basic advantages of using it with the industry 4.0 environment.

- The first area is the financial transaction, from where this technology got its popularity. In the case of the supply chain, where cross-border transit and multiple currencies are involved, various issues related to currency exchange can be avoided.
- Information logging and tracing could be useful in analyzing and identifying quality-related faults. An example of this is after-sale recalls of products due to some quality issue. This can cause financial and reputational loss to the organization. With the help of blockchain technology, complete information about the product from its inception to its delivery can be stored on the blockchain. This information can be analyzed easily to find the root cause of the fault and can diminish the overall impact of the recall.
- The cryptographic power of blockchain technology can be used to secure the exchange of information over a public network. With the help of this technology, we can avoid unauthorized entities to view or manipulate data being exchanged.
- Blockchain technology can provide digital identities to authenticate entities of industrial networks such as machines, tools, etc. These identities can be used to track the transaction performed by any organization, so that non-repudiation can be achieved. These identities will be unique just like the identities provided by government organizations.

- Smart Contract [25] is another benefit that can be achieved using blockchain. It is an automated agreement between communicating organizations that can be triggered on the occurrence of some event without the intervention of a third party. These are very useful where organizations do not trust each other.

As per the cost of a data breach report 2019 by IBM [24], automation has a great impact on security and can decrease the total cost of a data breach by a good margin. According to this report, the cost of the data breach was around USD 2.65 million if companies are employing a fully automated security solution. Organizations that are not using automated security solutions faced costs around USD 5.16 million, which is almost double that of fully automated security.

As per the cost of a data breach report 2019 of IBM [24], as far as the industrial sector is concerned, only 16% of industries have fully deployed automated security techniques, which is very few. Around 27% of industries have partially deployed automated security techniques, and 57% of industries are still not able to deploy automated security techniques, which is a very large part of the total industrial sector.

Blockchain technology can play a very crucial role in the automation of security in the industrial domain. There are various modules where this emerging technology can help. Some of the modules are supply chain, record management, IoT management [26], digital lifecycle management, etc. Some of the use cases of blockchain technology in these modules are given in Figure 12.6.

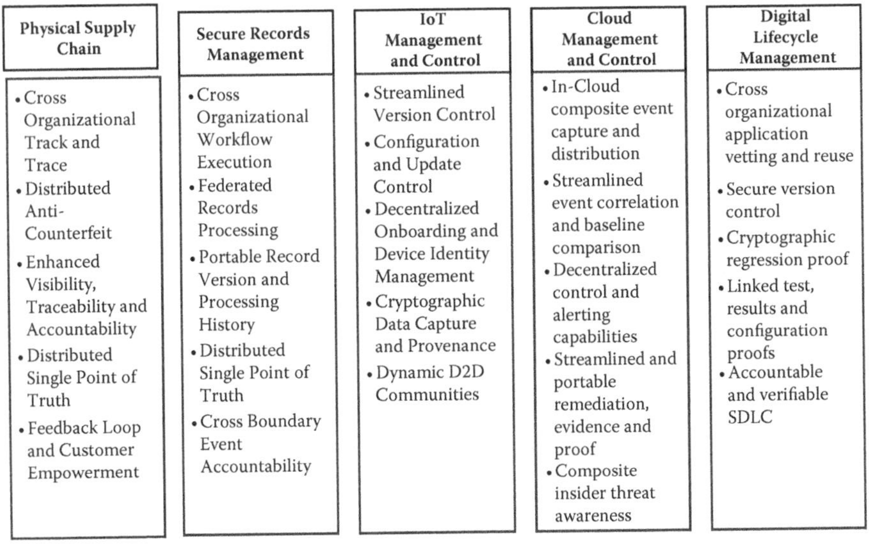

Physical Supply Chain	Secure Records Management	IoT Management and Control	Cloud Management and Control	Digital Lifecycle Management
• Cross Organizational Track and Trace • Distributed Anti-Counterfeit • Enhanced Visibility, Traceability and Accountability • Distributed Single Point of Truth • Feedback Loop and Customer Empowerment	• Cross Organizational Workflow Execution • Federated Records Processing • Portable Record Version and Processing History • Distributed Single Point of Truth • Cross Boundary Event Accountability	• Streamlined Version Control • Configuration and Update Control • Decentralized Onboarding and Device Identity Management • Cryptographic Data Capture and Provenance • Dynamic D2D Communities	• In-Cloud composite event capture and distribution • Streamlined event correlation and baseline comparison • Decentralized control and alerting capabilities • Streamlined and portable remediation, evidence and proof • Composite insider threat awareness	• Cross organizational application vetting and reuse • Secure version control • Cryptographic regression proof • Linked test, results and configuration proofs • Accountable and verifiable SDLC

FIGURE 12.6   Blockchain use cases in industry 4.0.

12.6 CONCLUSION

We are witness to the fourth industrial revolution and the growth of emerging technologies in the current digital era. The amalgamation of these emerging technologies with the industrial revolution can open a new dimension for a more secure framework. Blockchain technology is the new security protocol nowadays. It can be the new support for the industrial revolution as it can bring security, trust, and provenance to current industrial services. Many blockchain-based projects are being developed in recent times, but still, the exploration of blockchain technology is in its initial stages. Industry-centric investigation has yet to be explored and can be directed to resolve issues like scalability, integration, and implementation cost, etc.

REFERENCES

1. V. Chamola, T. Alladi, "Blockchain Applications for Industry 4.0 and Industrial IoT: A Review," *IEEE Access*, vol. 7, pp. 1–18, 2019.
2. C. Kim, M. Hassan Onik, "Personal Data Privacy Challenges of the fourth Industrial Revolution," in *ICACT 19*, 2019.
3. N. Paunkoska, N. Marina, J. Karamacoski, "Blockchain for Reliable and Secure Distributed Communication Channel," in *IAICT*, 2019.
4. O. Dluhopolskyi, S. Banakh, N. Moskaliuk, A. Farion, "Using blockchain Technology for Boost Cyber Security," in *ACIT*, 2019.
5. T. M. Fernandez Carames, P. F. Lamas, "A Review on Blockchain Technologies for an Advanced and Cyber-Resilient Automotive Industry," *IEEE Access*, vol. 7, pp. 17578–17598, 2019.
6. P. Fraga-Lamas, T. M. F. Carames, "A Review on the Application of Blockchain to the Next Generation of Cybersecure Industry 4.0 Smart Factories," *IEEE Access*, vol. 7, pp. 45201–45218, 2019.
7. L. Kong, J. Huang, "Towards Secure Industrial IoT: Blockchain System with Credit-Based Consensus Mechanism," *IEEE Access*, vol. 15, no. 6, pp. 1–10, 2019.
8. J. Al.-Jaroodi, N. Mohamed, "Applying Blockchain in Industry 4.0 Applications," in CCWC, USA, 2019.
9. S. Menon, N. S. Nair, R. Kumar, "Blockchain Solutions for Security Threats in Smart Industries," in ICCMC, India, 2020.
10. A. Beckmann, P. Kumar, A. J. M..Milne, "Cyber-Physical Trust Systems Driven by Blockchain," *IEEE Access*, vol. 8, pp. 66426–66437, 2020.
11. S. Tanwar, K. Parekh, S. Tyagi, N. Kumar, U. Bodkhe, "Blockchain for Industry 4.0: A Comprehensive Review," *IEEE Access*, vol. 4, pp. 1–37, 2016.
12. A. Zahid. M. Farooq, A. G. Khan, "A journey of WEB and Blockchain towards the industry 4.0: An Overview," in ICIC, 2019.
13. O. Schelen, K. Anderson, A. A. Monrat, "A Survey of Blockchain from the Perspectives of Applications, Challenges and Opportunities," *IEEE Access*, vol. 7, pp. 117134–117151, 2019.
14. T. Dasaklis, C. Patsakis, F. Casino, "A systematic literature review of blockchain-based applications: Current Status, Classification and Open issues," *Elsevier Telematics and Informatics*, vol. 36, pp. 55–81, 2019.
15. N. Mohamed, J. Al-Jaroodi, "Blockchain in Industries: A Survey," *IEEE Access*, vol. 7, pp. 36500–36515, 2019.
16. G. Zheng, K. Wong, Y. Zhu, "Blockchain Empowered Decentralized Storage in Air-to-Ground Industrial Networks," *IEEE Transactions on Industrial Informatics*, vol. 15, no. 6, pp. 3593–3601, 2019.

17. R. Xue, L. Liu, R. Zhang, "Security and Privacy on Blockchain," *CM Computing Surveys*, vol. 52, no. 3, pp. 1–34, 2019.
18. M. Zolanvari, A. Erbad, R. Jain, T. Salman, "Security Services Using Blockchain: A State of the Art Survey," *IEEE Communications Surveys & Tutorials*, vol. 21, no. 1, pp. 858–880, 2019.
19. S. K. Rana, S. K. Rana, "Blockchain Based Business Model for Digital Assets Management in Trustless Collaborative Environment," *Journal of Critical Reviews*, vol. 7, no. 19, pp. 738–750, 2020.
20. A. K. Rana, S. Sharma, "Contiki Cooja Security Solution (CCSS) with IPv6 Routing Protocol for Low-Power and Lossy Networks (RPL) in Internet of Things Applications," In *Mobile Radio Communications and 5G Networks* (pp. 251–259). Springer, Singapore, 2021.
21. A. O. Salau, N. Marriwala, M. Athaee, "Data Security in Wireless Sensor Networks: Attacks and Countermeasures," In: Marriwala N., Tripathi C. C., Kumar D., Jain S. (eds.) *Mobile Radio Communications and 5G Networks. Lecture Notes in Networks and Systems*, vol 140. Springer, Singapore, 2021. DOI: 10.1007/978-981-15-7130-5_13
22. A. Kumar, A. O. Salau, S. Gupta, K. Paliwal, "Recent trends in iot and its requisition with iot built engineering: a review," In *Lecture Notes in Electrical Engineering*, vol. 526. Springer, Singapore, 2019. DOI: 10.1007/978-981-13-2553-3_2
23. P. Dalal, G. Aggarwal, S. Tejasvee, Internet of Things (IoT) in Healthcare System: IA3 (Idea, Architecture, Advantages and Applications), 2020. Available at SSRN 3566282.
24. A. K. Rana, S. Sharma, "Industry 4.0 Manufacturing Based on IoT, Cloud Computing, and Big Data: Manufacturing Purpose Scenario," In *Advances in Communication and Computational Technology* (pp. 1109–1119). Springer, Singapore.
25. D. Lo, P. S. Kochhar, W. Zou, "Smart Contract Development: Challenges and Opportunities," *IEEE Transactions*, pp. 1–20, 2019.
26. Arun Kumar Rana[*] and Sharad Sharma, "Internet of Things Based Stable Increased-Throughput Multi-hop Protocol for Link Efficiency (IoT-SIMPLE) for Health Monitoring in Wireless Body Area Networks", *International Journal of Sensors, Wireless Communications and Control* vol. 1 (2021) 11:1. https://doi.org/10.21 74/2210327911666210120125154

13 Sensor Networks and Internet of Things in Agri-Food

Moses Oluwafemi Onibonoje
Afe Babalola University

13.1 INTRODUCTION

Agriculture and food production is steadily being revolutionized by the concept of sensor network and the Internet of Things (IoT) during the Fourth Industrial Revolution (Industry 4.0/4IR). In ensuring food security by nations and governments, efficient monitoring and control in the entire food chain are inevitable. The efficiency in the involving processes is ensured through the integration of various autonomous smart devices and the Internet, to achieve the desired objectives. Integrated sensor networks aid the aggregation of the plus capacity of each sensor node within the monitored field, and IoT incorporates the network with the Internet in helping the farmers and investors to improve their productivity [1]. The sensor networks and IoT are crucial composite concepts for the food revolution (Food 4.0) during the Industry 4.0, the descriptive relevance is shown in Figure 13.1. In food production through agriculture, IoT applications allow farmers to boost the quantity, quality, cost-effectiveness, and sustainability of their products. The concept allows the producers to determine the seeds, the quantity of applicable fertilizer, the possibly best harvest time, and the probable crop outputs. Also, IoT presents the tools that are suitable for detecting soil moisture, level of livestock feed, crop growth, and others. It enables food disruption through effective remote monitoring, management, and control of irrigation equipment and harvesters using digital sensors and the Internet.

Current works show that IoT comprises a network of sensors connected with the Internet, and with the capability of producing data that could be analyzed. With the combination of other synergistic technologies, such as machine learning and big data, IoT is predicted to significantly influence market analysis, production, and logistics in Food 4.0. Sensor technology is applied to collect data on animals, soil, and crops through the mishmash of machines, equipment, aircraft, satellites, and drones. Remote sensing technology has enabled the integration of these component units into the entire food value chain: farming production, supply chain, and other post-harvest systems such as weather data provision and product processing. IoT incorporates local and remote sensing units for collecting a large range of production

DOI: 10.1201/9781003140443-13

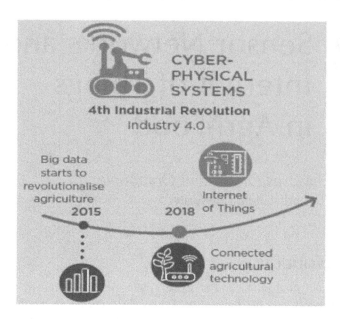

FIGURE 13.1   The relevance of the IoT to the era of cyber-physical systems.

management data. The systems are being integrated with effective data analytics for providing tools necessary for decision guiding at a good resolution. IoT will provide the complete smartness needed in the applications of low-cost sensor systems and Internet in the entire food lines, as shown in Figure 13.2 [2]. Smart food systems use smart devices, increase automated processes, improve energy efficiency, and boost the application of information technology.

The agriculture and food sector have been the major contributors to the economies of nations and the quality of their people's lives. The various crops and grains being produced, continuously need to be effectively monitored and stored for consumption, marketing, and other purposes. This ensures value addition from the products and also supports food security in those places. The network of sensors and IoT are the crucial concepts for providing the needed platform for efficient monitoring and control in the storage and management of the crops [3,4]. Recently, innovations and inventions in IoT systems have integrated real-time frameworks to solving various problems. The available resources for IoT solutions are limited, but there is a wide range of services being provided by its devices. IoT incorporates the sensing, control, and communication capacities of the different smart devices with the Internet to arrive at more robust research solutions [5–7].

The emerging tendency of cloud computing and IoT has transformed ordinary devices and environments into "smart things." The basic idea in cloud computing is the availability of shared aggregation of resources on-demand, rather than the previous dependence on the local server for data access to run an application. Cloud computing is combining with IoT to provide a new paradigm for creating business models [8]. IoT depicts a concept whereby physical devices ("things") are integrated with in-built communication capacity among the networked devices and

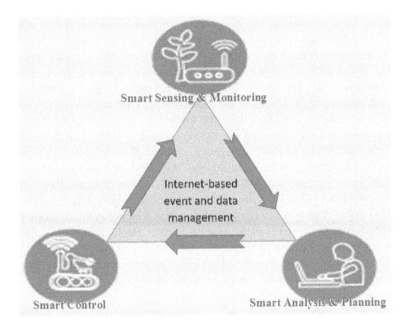

FIGURE 13.2 Smart architecture of IoT.

the environment. The concept represents a phenomenon with the ability to re-construct the technological norms, and also automate the operations and actions within the networks. Also, IoT encompasses multi-faceted approaches with the involvement of practicality in implementation, with a considerable trend in information technology [9,10].

The concept of IoT is exemplified in the implementation of various smart systems: smart house, smart city, smart enterprise, smart farm, smart environment, smart transport, smart wearable, and others [9,11–13]. A lot of research works are being done in the implementation of sensor networks and IoT systems with applications in the food chain. Different drawbacks are being identified and resolved by many researchers and academia. Most times, the limitations in the various systems are being resolved through the selection of the possible components with comparative advantages and incorporation of parameters trade-off [14,15]. The distributed system in [3] incorporates multi-networked wireless sensors with the Internet to provide efficient monitoring and control for a mini grain storage facility. Many identified weaknesses of the previously designed grain management units were resolved in the implementation of the distributed system. Other studies present possible mathematical modeling solutions to the issues in the sensor network systems for food monitoring, intending to develop and optimize the models to achieve optimal trade-offs among the multi parameters. Some IoT platforms employed in the food industry allow producers to manage a large amount of data being relayed and collected through sensors, connected devices, and cloud services. Therefore, IoT often forms a strong synergy with big data analytics in a lot of applications in the food sector.

Wireless sensor network (WSN) is a widely recognized technology and is important for the development of the food sector of the future. Industries and academia have focused on implementing various applications of the concept. The network is always an aggregation of the sensing and communication capabilities of the low-power, low-cost nodes for specific functions. WSN and IoT have a very good synergy with other concepts of the fourth industrial and food revolution, with a wide range of applicability including, but not limited to, environmental monitoring, process control, and solving other diverse problems [3], [16], [17]. A typical example is a work that presents an advisory system approach aimed at providing undergraduate students with the basic application of WSN in railway level crossings [18], while another presents an IoT-based approach in conditioning and controlling a storage room in real-time [19]. Some other works also proposed the possible WSN application in fire detection and control. The various techniques have shown a quite moderately larger scale of research range than other fields. Nonetheless, the area that has not been adequately implemented and justified is the monitoring, mitigation, and notifications of the early development of fire in real time. Fire outbreaks in farmlands and storage facilities can cause an unimaginable food loss and possible loss of lives and other valuables [20–22].

The research areas in WSN and IoT can be categorized into three: component level, system level, and application level. The focus of the component level research is about the improvement of the sensing, communication, and computation capabilities of the sensing unit. System-level research focuses on the routing mechanism of the coordinating and networked sensors to attain scalability and energy efficiency. Research at the application level focuses on the sensor data processing in line with the objective of the system. Therefore, the highlight of the WSN and IoT systems is best described in a generic, modular, and interoperable mode. Regardless of the application area of a WSN and/or IoT system, the basic constituent units are comparatively the same. The consisting units of a WSN and IoT system include sensing unit, micro-processing unit, radio, and connectivity unit, software unit, Internet, and others. The system architecture of a typical IoT- and WSN-based system is as shown in Figure 13.3.

13.2 SUMMARY

The concept and framework of the sensor network and IoT, even in the food revolution, can be adequately described in three levels: component level, system level, and application level.

13.2.1 COMPONENT LEVEL

The first important component of the IoT/WSN system is the sensing unit. Sensor nodes form the heart of the network, and intermittently detect the parameter, process the data, and send it to the sink node within the communication network. The sink node represents the aggregation point for the data. It receives, processes, and stores data, and route the data to the coordinating node through the Internet or other wireless communication. The coordinating node sets up the network and acts as the

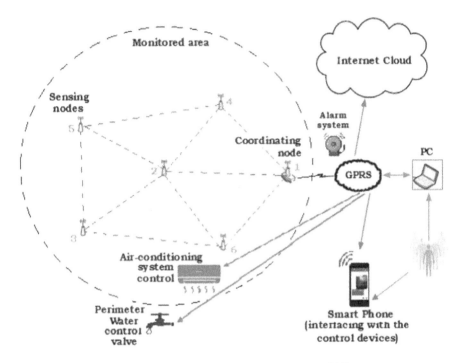

FIGURE 13.3 An architecture of a typical IoT – WSN system [19].

central point for the sensor nodes. It also acts as the gateway to accessing other networks, and also serves as the access point for human interface.

13.2.1.1 Sensing Unit

The sensing unit within the WSN/IoT systems consists of a sensor and analog-to-digital (ADC) converter. A sensor can be described as a smart tiny component developed based on sensor technology, with fast data acquisition and low power signal processing capabilities, having relative energy constraints, good accuracy and reliability, and less real-time maintenance. A sensor mainly detects events, processes the data locally, and transmits it. The transmission of the data is initiated by the unit based on the routing algorithm being used. The sensor(s) collaborate with other components within the sensing node such as microcontroller, processor, memory, transceiver, and power source. The interaction of the sensors with the environment is measured and transferred to the microcontroller. The sensing nodes collaborate to fulfill their tasks by using wireless links. Meanwhile, the nodes are configured to assign their functionality. The coordinating node has a different configuration from the sensing node. The configuration is implemented with defined software architecture.

13.2.1.2 Data Processing Unit

The processing unit is virtually the most complex composite of the WSN/IoT system. It consists of the microcontroller, memory, and the operating system.

The microcontroller performs the computational tasks on the received data, monitors and controls the battery power, and calculates the next hop to the coordinating node. The flash memory is responsible for the storage of program code and the computation of the procedures by non-volatile read access memory. The operating systems in sensor nodes are less complex, such as LiteOS, TinyOS, Contiki, and RIOT.

13.2.1.3   Communication Unit

The sensor nodes and the coordinating node exchange messages within the network via the communication unit. The unit receives query and command from the microcontroller unit (MCU) and transmits data. A transceiver is the main constituent of the unit, and responsible for receiving instructions from the processing unit and communicating with the other networked nodes through the antenna. The transceivers of most sensor nodes communicate within the allowable free licensed frequencies 173, 433, 868, and 915 MHz; and 2.4 GHz, with the electromagnetic waves media for WSN [23]. The transceivers are configured to execute the topology and interconnectivity designed for the nodes.

13.2.1.4   Power Sources

Each of the units and components within the network nodes requires energy to perform the designed tasks. The power generating unit for the nodes is designed in two crucial forms of storing energy and providing in the required form and consuming energy through scavenging. The unit has two main components: battery and DC-DC converter. The battery powers up the sensor nodes and discreetly determines how long the network functions, based on whether the battery is rechargeable or not, while the DC-DC converters help to regulate the node voltage level. The lifetime of the network depends largely on the lifetime of the power source. The power consumption of each of the sensor nodes, hence the network can be optimized by configuring the communicating transceiver in a cyclic sleeping mode, modifying the MCU to operate in a power-down mode, and removing the indicating lights. As reported by [3], a sensor node is powered by a 9 V battery of 1200 mAh, having a modified Arduino MCU consumes 0.045 mA, with transceiver transmit current of 45 mA was a lifetime of about 27 h. When the sensor node has a sleeping configured transceivers with data transmitting five times every 4 h, the lifetime of the node is increased to approximately 3158 days.

13.2.2   System Level

The IoT design and framework require a combination of multiple blocks such as custom components for energy-efficient control, computational mapping in the central processing unit (CPU), integrating the custom hardware components, the CPUs, sensor/actuator unit, and the communication peripherals [24,25]. In the design effort, if only one portion of the process is accelerated, a substantial part of the design and integration efforts for the other components may be left out. A system-level design process is required in every sensor network and IoT system. The process incorporates hardware/software co-design, system integration,

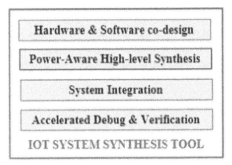

FIGURE 13.4 Synthesis tool for an IoT System [24].

compute-component design, prototyping, and verification. A typical synthesis tool for an IoT system is as shown in Figure 13.4.

The limited resources available in the network of wireless sensor nodes require efficient utilization. Precise modeling of the integrated sensor node units is necessary for the overview of accurate system-level design decisions on the architecture for applications and real-time operating systems.

13.2.3 APPLICATION LEVEL

The demands of coping with the complexity and dynamic agro-food environment require that the hardware devices be fully upgraded for compliance in universality, expansibility, intelligence, reliability, and endurance while minimizing the operative challenges and costs [26]. Also, the local network is required to be shielded from interference from other networks. The filtering, interoperability, and semantic annotation of the generated data by the different IoT devices need to be realized adequately. With this, the application-level big data decision support is effectively optimized through the use of a huge amount of heterogeneous data.

In adopting the IoT system, anonymity, security, and control over the information access rights are very important. The impact of the WSN/IoT systems on the social economy and agro-ecological environment must be considered in their applications in agro-food, to realize sustainable food development.

13.2.4 DESIGN ISSUES AND CHALLENGES

Efficient performance of WSN/IoT systems can be achieved by overcoming some essential issues and design constraints. In designing the routing protocol, consideration must be given to the application areas, design constraints, and network architecture. The constraints and challenges to the design issues are very important for observations.

13.2.4.1 Faults Tolerance and Adaptability

A section of the network of wireless nodes can be observed to be inactive whenever the nodes there fail or are blocked. The node failure is often caused by the unavailability of power and transmission or interference-related problems. The overall change of network configuration or changing of connectivity through efficient routing protocols is a major tool for ensuring sensor network adaptability in the case of a fault. The reliability of any sensor network can be modeled by using a Poisson distribution to analyze the probability of failure in the network [27]. The probability of no failure within a time interval (0, t)

$$R_t(t) = e^{-\lambda k t} \tag{13.1}$$

where λk is the related failure rate of each sensor k, the period interval is t.

13.2.4.2 Network Topology Issue

Sensor network topology represents the interconnectivity relationship among the sensor nodes. Many of the characteristics of the network are affected by the topology, such as data processing and routing, network capacity, latency, and others. The maintenance and control in the network topology are classified into three stages [28]: pre-deployment and development, post-deployment, and additional sensor redeployment.

13.2.4.3 Energy Consumption with Network Accuracy

Energy constraints are major issues in the sensor network and IoT systems. Mostly, such networks depend solely on batteries (typically, <0.5 Ah 1.2V). Inevitably, power conservation and management, also energy harvesting becomes crucial. Sensor networks and IoT require routing protocols that are capable of efficiently managing energy consumption without losing the accuracy of the networks. The estimation of the node battery life helps determine the feasibility of the network protocol to give an application-based solution.

13.2.4.4 Coverage Efficiency and Connectivity

The coverage efficiency of the system reflects how adequate the area of interest is tracked by the deployed sensors. The coverage algorithm influences the sensing of the data by the end nodes within the region and transmitted to the coordinating node using the routing algorithm. In [3], the interest area W_n within the coverage of a set of sensor S_k was defined as 2-dimensional space. The sensors have their location well-defined. The work aimed at ensuring that every point P_i fell into the sensing range of the sensors.

$$Z_i \in W_n = \{(x_1, y_1), (x_2, y_2), \dots\dots(x_6, y_6)\} \tag{13.2}$$

$$S_k = \{s_1, s_2, \dots\dots s_6\} \tag{13.3}$$

The union of the individual node's coverage equals the actual coverage area for the entire network nodes. The coverage efficiency was then evaluated as the ratio of the union of each node's coverage to their sum.

Connectivity highlights the efficient routing of sensed data from the end nodes to the coordinating node. The radio interconnectivity among the nodes influences the design of the routing protocol and the dissemination techniques [28]. The inter-relationship between the coverage and connectivity is the scheduling, a major component for improving the sensor lifetime and avoiding data replication.

13.2.4.5 Scalability in Architecture

A sensor network has a routing protocol requirement of working with a huge number of nodes, and capable of accepting more sensor nodes into the existing network through the coordinating node. The coverage efficiency of the network, as well as the reliability and accuracy of the data processing, is affected by the density of the sensor nodes. The density of sensor nodes within a region can be calculated using the work of [27].

$$\mu_R = \frac{N\pi R^2}{A} \tag{13.4}$$

where A represents the sensor field, N represents the number of sensors within the field, and R represents the transmission range of the radio. μ_R is the density of the sensors in the transmission radius within the region.

13.2.4.6 Data Aggregation

Data aggregation describes the routing techniques for fusing the various data from the different sensor nodes into useful information and eliminating replicas. Bandwidth utilization, traffic optimization, and energy efficiency depend largely on the techniques of data aggregation.

13.2.4.7 Data Delivery Modes

The delivery of data to the coordinating node is dependent on the application model of the sensor network. The models can be categorized into event-driven, time-driven, query-driven, or a combination of two or more. The efficiency of the routing algorithm is largely influenced by the delivery model of the data via energy consumption, route stability, and reliability [29].

13.2.4.8 Self-Configuration

Sensor networks are deployed in dynamic environments where the constituent sensor nodes are unattended. A major feature is the ability of the nodes to self-configure themselves by establishing a topology that supports maximum throughput, minimum delivery delay, and minimum energy consumption [30].

13.2.4.9 Safety and Security

Network safety is a major constraint in the design and application of IoT and sensor network systems. The new technology is susceptible to hackers' invasion due to the component sensor nodes without a particular identification number. Therefore, the systems require a distributed approach to security issues through confidentiality, authorization, authentication, non-repudiation, and integrity.

13.2.4.10   Quality of Service (QoS)

Sensor networks and IoT systems often manage a trade-off between energy consumption and quality of data. The design of the systems must satisfy the QoS metrics in the form of error rate, bandwidth, data latency, packet loss, and others. The network should provide the needed service to selected network traffic across various technologies. The data latency explains the total duration for a network to transmit data from the sensing node to the sink. However, some routing protocols generate excessive overheads in the implementation algorithms. These are constraints to overcome in the design of the systems.

13.2.4.11   Hardware Constraints

The various units constituting the nodes within the sensor networks and IoT have the requirements of collective integration for the successful implementation, deployment, and application of the network systems.

13.2.4.12   Optimization for Robustness

The various sensor networks and IoT require various techniques to guarantee system robustness. These may include the optimization of the power consumption, discovery, and repair in the communication route, detection, and correction of errors in the network systems [3].

13.2.4.13   Production Costs

The overall cost of designing a sensor network system is the total cost for each node and the integrating network. Most networks consist of a large number of sensing nodes. If the cost of producing each node is high, there is an obvious concern for a very high cost of developing the system with an attendant prohibitive and impeding adoption for the spread of sensor technology. Hence, the cost of each node needs to be kept as low as possible.

13.2.5   CASE STUDY: SENSOR NETWORKS AND IoT IN THE FOOD CHAIN

Sensor networks and IoT have employed sensor and smart technologies to ensure food security, by ensuring reliable food systems across the entire spectrum of the food chain. Food security includes the involvement of the food systems in ensuring enough food for everyone to live healthily. The stages of the food chain include food production, food harvesting, food transport and distribution, and management of the bio-waste outputs. Sensor technology and IoT are employed in the monitoring and control of the entire food cycle. IoT is changing the world view about food, ranging from food deliveries, better management of crops, and data collection process for better business decisions.

The traditional method of food production is adapting to new technologies. For instance, in Italy, people could rent plots of land in remote areas, while directing the cultivation of the crops through the Internet at their locations. The different uses of the Internet do share the costs and update their sites, capable of directing the rural farmers to cultivate interested crops according to the demands of the land renters.

The food revolution through sensor technology and IoT has been tremendous. Consumers now think differently about food and its various processes. IoT has impacted food in various ways that include: excellent quality assurance using smart thermostats, warehouse management through inventory sensors, inventory reminders and food alerts using smart appliances, effective crop management using drones, self-driving tractors, and others.

13.2.5.1 Food Quality Assurance Through Smart Thermostats

A smart thermostat is developed for food companies to monitor products' temperature in real-time and initiate a control measure for safe public consumption [31]. During food supply and transport, there is a high tendency for temperature-sensitive products to be warm, thereby increasing the risk of food-borne diseases and an impact on public health. Many companies now invest in temperature-controlled packaging through smart thermostats for real-time monitoring. Whenever the remotely communicated temperature of any food product is below a threshold, companies take the recall of the product to ensure consumers' safety. Also, consumers can scan the QR codes on food products in the stores to verify the safe condition for the products' consumption.

13.2.5.2 Waspmote Devices in Food Production, Distribution, Supply, and Waste Management

Sensor technology and IoT have been employed in developing Waspmote devices to improve quality of life through food, from the production stage to the consumption stage and the garbage disposal stage. Improvements have been enabled for the consumers to be involved in the entire food cycle, and for the governments and authorities to utilize the scarce resources in ensuring food safety for all inhabitants [32]. Inventions from innovative models have improved food production. Waspmote has built networks of various sensor devices with little resources and capability of monitoring crop production across the entire food cycle. A typical sensor device from Waspmote is as shown in Figure 13.5(a). The sensor device can be integrated into different applications.

The combination of temperature, humidity, and light sensors could be applied to detect the risk of frost. Data from soil humidity measurements are used in the monitoring process to prevent inherent plant diseases, also in the management of watering requirements. This aids in the control of nurseries conditions, and close monitoring of delicate crops and high-performance vineyard fruits, in which the slightest climatic change does affect the outcomes. Also, optimum conditions for the individual crop can be determined and the obtained figures compared for the best harvests possible. The networks of these wireless nodes can be integrated into the monitoring of the isolated areas with difficult accessibility, where special crops like truffles and mushrooms do grow.

Multiple parameters from various environmental sources, with a wide application range, can be monitored using the Waspmote Agriculture board as shown in Figure 13.5(b). The board is very useful in the growing development analysis and

(a)   (b)

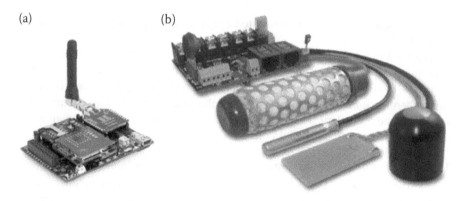

FIGURE 13.5 Wireless devices from Waspmote: (a) a sensor device, (b) an agriculture board [32].

weather observations. It has the capacity for integrating with different applicable sensors. The system consisting of the board can be used to reduce on-site labor requirement and resource intensity during the food growth phase, through advanced defined parameters for each crop. The crop cultivation system is automated using the sensor readings at any desired stage of the food production stage. The system also can turn off watering or switch over water supply through a control provided from an SMS or the ZigBee network. Hence, effective water management.

In the food distribution stage, the probability of food damage during storage and transport is high. Food safety issues and traceability are trending research areas of interest. The supply-chain traceability of food is possible through an IoT-based system [26]. Real-time control of the merchandise's condition at any time is possible through integrated technology in Waspmote sensors, GPS, and clocks. During food product transport, environmental samples can be detected and stored. The data helps to track the condition of the product at any point in time, whether there has been exposure to undeserved damp or high temperatures. Food contamination, unauthorized openings, impact suffer, and others are also part of the things that can be detected along with product transport. The events board with various sensors is as shown in Figure 13.6(a). GPS and GPRS/3G modules are integrated into the event boards, which help in food traceability and provision of location details as shown in Figure 13.6(b).

IoT devices are also developed for the food supply stage, for instance, in the supermarket. Operating smart labels on the merchandise containers through the transmission of wireless food information across the ZigBee network is made possible. The RFID (Radio Frequency Identification) in the developed system enables the detection of expired food, and even alerts the consumers on the inherent danger of eating a particular food product with specific unsuitability to his health allergies. Also, IoT sensors can be deployed for monitoring in public and risk disposal areas, to keep the contamination levels within safe levels.

(a) (b)

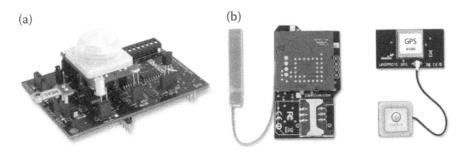

FIGURE 13.6 IoT devices in the food distribution stage, (a) Events board with sensors, and (b) GPS and GPRS/3G modules [32].

13.2.5.3 Wireless Nodes in Granary Monitoring

Developed wireless sensor nodes and IoT systems are applied in a modular storage facility useful for middle-level marketers and small-scale farmers for monitoring stored grains [3]. The system aids in ensuring food security and good returns on quality grains, as a result of the well-monitored and controlled storage bins. The developed nodes are as shown in Figure 13.7.

13.2.5.4 Milk Production in Voshazhnikovo Farm Increased from 28 to 35 Liters Per Cow and Day Using IoT and Machine Learning

The second point in the sustainable development goals (SDG) is zero hunger. Eliminating global hunger in the face of the surging population is making food production as very effective as possible. Russia is revolutionizing its food industry through the synergistic technologies of Industry 4.0: machine learning, the IoT, and genetically modified husbandry. The huge milk deficit in the country [33] has activated the process of using genetically modified cattle breeding. ALAN-IT, a Russian company, developed a management system for dairy production [34] as is illustrated in Figure 13.8. The system is supported with cloud-based analytics, to aid decision-making on livestock management through the recorded environmental parameters such as humidity, temperature, cow health, and pressure.

The forecasting performance of the system involves the collection of data with external sources and sensors, generating excel reports to the cloud-based data platform on the farm. LoRaWAN serves as the communication protocol connecting devices to the gateway, and the gateway to the cloud. The system was implemented in a Voshazhnikovo farm [35] with an 8,000-head cattle capacity in which 4,500 are dairy cows. Sequel to the implementation of the system, 125 tons of milk (28 liters per cow per day) were being produced. When implemented, the system found a correlation between the parameters such as nutrition, temperature, and even daily farm-workers' performance. For instance, the system predicted that if the temperature decreased, the feed needs for the cow increased. The staff was then warned about the changes through SMS and mail notifications. With adequate response to the warning and proper feeding, the milk production increased. Three months after the implementation, the milk yields have recorded an 18% increase to 35-liters per cow per day.

FIGURE 13.7 A WSN-IoT granary monitoring system (a) sensing node, (b) coordinating node, and (c) system nodes [3].

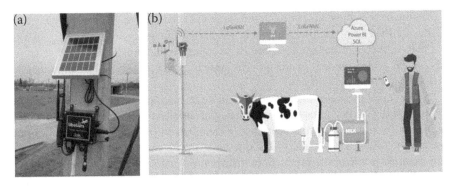

FIGURE 13.8 Smart diary monitoring system, (a) device unit for temperature recording, (b) system installation diagram [34].

13.2.5.5 An IoT-based Agro-Tech Measurement System in Aragon, Spain

The main objective of the system is to provide an autonomous measurement of different variables in the deployed farm and be sent to the cloud in which control actions are initiated from the Internet cloud based on the measured parameters [36]. The measurement system is as shown in Figure 13.8. The nodes are placed on sites while they communicate with the system gateway using the 4G, which, in turn,

FIGURE 13.9 Communication diagram for IoT-based agrotech project in Spain [36].

sends the data to the cloud through the same 4G. The system gateway is located in Zaragoza for easy system management for the developers.

The condition of the device could be managed in the cloud and the information should be accessible and available. The agriculturist can access real-time data to make better strategic decisions.

Summarily, there are many variables to keep track of in the food chain of the Industry 4.0. Food manufacturers, farmers, and suppliers will need the adoption and adaptability of every need concept and technology to maximize their outputs and ensure food security and safety. The effects of the changing climatic condition and any minimal error in the production, distribution, and supplies can result in grave consequences on the company image, food prices, and consumers' health. The smart and digital tools of sensor networks and IoT are veritable resources for the interventions (Figure 13.9).

REFERENCES

1. J. P. Tomás, "Three Precision Agriculture IoT Case Studies," 2017. [Online]. Available: https://enterpriseiotinsights.com/20170516/smart-farm/20170516smart-farmthree-precision-agriculture-iot-case-studies-tag23-tag99. [Accessed: 26-Sep-2019].
2. University of Stellenburg Business School, "The Future of the Western Cape Agricultural Sector in the Context of the 4th Industrial Revolution, Review: Agriculture in 4IR & Its Drivers – A Global Perspective," Cape Town, 2017.
3. M. O. Onibonoje, N. I. Nwulu, and P. N. Bokoro, "A Wireless Sensor Network System for Monitoring Environmental Factors Affecting Bulk Grains Storability," *J. Food Process Eng.*, vol. 42, no. 7, pp. 1–13, 2019.
4. D. Jayas, J. Paliwal, and N. Visen, "Automation and Emerging Technologies: Multi-Layer Neural Networks for Image Analysis of Agricultural Products.," *J. Agric. Eng. Res.*, vol. 77, no. 2, pp. 119–128, 2000.
5. J. Cheng, Y. Gao, N. Zhang, and H. Yang, "An Energy-Efficient Two-Stage Cooperative Routing Scheme in Wireless Multi-Hop Networks," *Sensors*, vol. 19, no. 5, p. 1002, 2019.
6. M. O. Onibonoje, N. I. Nwulu, and P. N. Bokoro, "An Internet-of-Things Design Approach to Real-Time Monitoring and Protection of a Residential Power System," in *2019 the 7th International Conference on Smart Energy Grid Engineering*, pp. 113–119, 2019.
7. S. Darroudi, R. Caldera-Sanchez, and C. Gomez, "Bluetooth Mesh Energy Consumption: A Model," *Sensors*, no. Special Issue, pp. 1–14, 2019.
8. P. K. Ghosh, M. Sahinoglu, and V. V. Phoha, "Resource Management for Uninterrupted Microgrid Operation," in *2019 the 7th International Conference on Smart Energy Grid Engineering*, pp. 321–324.
9. B. Mrazovac, M. Bjelica, N. Teslic, and I. Papp, "Towards Ubiquitous Smart Outlets for Safety and Energetic Efficiency of Home Electric Appliances," In *Consumer Electronics-Berlin (ICCE-Berlin) 2011 IEEE International Conference*, pp. 322–326, 2011.
10. D. Uckelmann, M. Harrison, and F. Michahelles, "An Architectural Approach Towards the Future Internet of Things," In Architecting the Internet of Things (pp. 1–24). Springer, Berlin, Heidelberg, 2011.
11. R. Y. Li, H. Li, Y. Ch., K. Mak, K. Ch., and T. B. Tang, "Sustainable Smart Home and Home Automation: Big Data Analytics Approach," *Int. J. Smart*, vol. 10, no. 8, pp. 177–198, 2016.

12. S. Li, H. Wang, T. Xu, and G. Zhou, "Application Study on Internet of Things in Environment Protection Field," *Lect. Notes Electr. Eng.*, vol. 133, pp. 99–106, 2011.

13. S. Zach and C. Bormann, *6LoWPAN: The Wireless Embedded Internet*. John Wiley & Sons, 2011.

14. M. O. Onibonoje, K. S. Alli, T. O. Olowu, M. A. Ogunlade, and A. O. Akinwumi, "Resourceful Selection-Based Design of Wireless Units for Granary Monitoring Systems," *ARPN J. Eng. Appl. Sci.*, vol. 11, no. 23, pp. 13754–13759, 2016.

15. M. O. Onibonoje, A. O. Ojo, and T. O. Ejidokun, "A Mathematical Modelling Approach for Optimal Trade-offs in a Wireless Sensor Network for a Granary Monitoring Systems," *Int. J. Technol.*, vol. 10, no. 2, pp. 212–218, 2019.

16. A. R. V., R. R. Raj, and N. K. Prakash, "Industrial Automation Using Wireless Sensor Network," *Indian J. Sci. Technol.*, vol. 9, no. 11, pp. 1–8, 2016.

17. M. O. Onibonoje, U. N. Ume, and L. O. Kehinde, "Development of an Arduino-Based Trainer for Building a Wireless Sensor Network in an Undergraduate Teaching Laboratory," *Int. J. Electr. Electron. Sci.*, vol. 2, no. 3, pp. 64–73, 2015.

18. M. O. Onibonoje, "Development of a Level-Crossing Advisory Prototype Using Wireless Sensor," *ABUAD J. Eng. Res. Dev.*, vol. 2, no. 1, pp. 154–160, 2019.

19. M. O. Onibonoje, N. I. Nwulu, P. N. Bokoro, and S. L. Gbadamosi, "An IoT-Based Approach to Real-Time Conditioning and Control in a Server Room," in *IEEE IDAP*, pp. 1–6, 2019.

20. A. Sardouk, M. Mansouri, L. Merghem-Boulahia, D. Gaiti, and R. Rahim-Amoud, "Crisis Management Using MAS-Based Wireless Sensor Networks," *Comput. Networks*, vol. 57, no. 1, pp. 29–45, 2013.

21. A. Clemente, J. Mart\'\inez-de Dios, and A. O. Baturone, "A wsn-Based Tool for Urban and Industrial Fire-Fighting," *Sensors*, vol. 12, no. 11, pp. 15009–15035, 2012.

22. A. Molina-Pico, D. Cuesta-Frau, A. Araujo, J. Alejandre, and A. Rozas, "Forest Monitoring and Wildland Early Fire Detection by a Hierarchical Wireless Sensor Network," *J. Sensors*, vol. 2016, pp. 1–8, 2016.

23. M. O. Onibonoje and T. O. Olowu, "Real-Time Remote Monitoring and Automated Control of Granary Environmental Factors Using Wireless Sensor Network," in *2017 IEEE International Conference on Power, Control, Signals and Instrumentation Engineering (ICPCSI)*, pp. 113–118, 2017.

24. K. M. Virk, K. Hansen, and J. Madsen, "System-Level Modeling of Wireless Integrated Sensor Networks," in *International Symposium on System-on-Chip (SoC) IEEE*, 2005.

25. L. Yang *et al.*, "System-Level Design Solutions: Enabling the IoT Explosion," in *2015 IEEE 11th International Conference on ASIC (ASICON)*, 2015.

26. X. Shi *et al.*, "State-of-the-Art Internet of Things in Protected Agriculture," *Sensors*, vol. 19, no. 1833, pp. 1–24, 2019.

27. S. Kalantary and S. Taghipour, "Survey on Architectures, Protocols, Applications, and Management in Wireless Sensor Networks," *J. Adv. Comput. Sci. Technol.*, vol. 3, no. 1, pp. 1–11, 2014.

28. S. Gowrishankar, *et al.*, "Issues in Wireless Sensor Networks," in *Proceedings of the World Congress on Engineering*, 2008.

29. Rajashree V. Biradar, *et al.*, "Multihop Routing in Self-Organizing Wireless Sensor Networks," *Int. J. Comput. Sci.*, vol. 8, no. 1, pp. 1694–0814, 2011.

30. A. Zaheer *et al.*, "A Survey of Wireless Sensor Network- Software Architecture Design Issues," *Int. J. Comput. Sci. Telecommun.*, vol. 3, no. 3, 2012.

31. D. Madden, "The Future of Food: 6 Ways IoT is Impacting Food Supply Chains," 2019. [Online]. Available: https://supplychainbeyond.com/category/technology/internet-of-things/. [Accessed: 25-Nov-2019].

32. A. Bielsa and M. Boyd, "Wireless Sensor Networks to Monitor Food Sustainability," 2012. [Online]. Available: http://www.libelium.com/food_sustainability_monitoring_sensor_network/. [Accessed: 30-Nov-2019].

33. FAO, "Analysis of Dairy Production," 2018. [Online]. Available: http://www.fao.org/dairy-production-products/production/en/. [Accessed: 02-Dec-2019].

34. Alan-IT, "Diary Production Analytics," 2019. [Online]. Available: https://smart4agro.ru/dpa.html. [Accessed: 03-Dec-2019].

35. Libelium, "Milk Production in Voshazhnikovo Farm Increased from 28 to 35 Litres Per Cow and Day Using IoT and Machine Learning," 2019. [Online]. Available: http://www.libelium.com/how-a-dairy-farm-increased-their-milk-production-18-with-iot-and-machine-learning/. [Accessed: 02-Dec-2019].

36. Libelium, "New Vineyard Project Developed with Libelium IoT Platform on Agrotech the App for Crop Management," 2019. [Online]. Available: http://www.libelium.com/new-vineyard-project-developed-with-libelium-iot-platform-on-agrotech-the-app-for-crop-management-powered-by-efor-and-ibercaja-on-microsoft-azure/. [Accessed: 04-Dec-2019].

14 Design and Development of Hybrid Algorithms to Improve Cyber Security and Provide Securing Data Using Image Steganography with Internet of Things

Abhishek Mehta[1] and Trupti Rathod[2]
[1]Research Solar at Department of Computer and Informative Science, Sabarmati University, Ahmadabad, Gujarat, India. & Assistant Professor Parul Institute of Computer Application, Parul University, Vadodara, Gujarat, India
[2]Assistant Professor, Vidyabharti Trust College of Master in Computer Application

14.1 OUTLINE

The Internet of Things (IoT) is one of the greatest advancements used in multiple operations as well as cutting-edge devices. It involves system-to-system communication, computer-generated data, and their associations. Based on a simple technology, the IoT framework can be implemented in household appliances and numerous other information technology devices. With IoT, individuals can send to their home appliances using smart apps or devices. Moreover, Cisco's Internet BSG had anticipated the number of IoT gadgets to be increased by twofold (50 billion gadgets) by 2020 [1]. The comfort and "insightfulness" of IoT gadgets help in the development of a number of such gadgets. Notwithstanding, a few people may in any case be hesitant to have these gadgets in their homes. In spite of the fact that IoT gadgets may make an individual's life much simpler, several security concerns have been already raised by data specialists (The Uncertainty of Belongings), which brand the IoT gadgets among the top five security dangers in 2016 [2].

DOI: 10.1201/9781003140443-14

In actuality, data divulgence and principle safety issues in IoT framework cannot be disregarded [3]. In addition, IoT gadgets with poor safety are difficult to secure e.g. IP photographic camera, smart television, and other home appliances. However, with varied designs and connection over the Internet, dynamic and aloof security could be easier to take care of using the IoT framework so as to secure the human resource data from being compromised in different organizations that make use of innovative devices. For instance, hackers have a more easy opportunity to block and mediate an information broadcast between IoT gadgets as it involves large data transfer, connected over the Net. The data that the programmers are seeking are identified by verification, installment, or even authoritative privileged insights. By getting validation data through snooping or capturing it from an association, the attacker can have control over IoT gadgets, utilizing the recovered data for misuse [4]. To address the issue of classification in the IoT framework, we propose security conspires that includes associations with the Net. Meanwhile, the IP photographic camera is painstaking as the IoT gadget in this examination, communicating the delicate data between IP cameras and a home worker that is dependent on utilizing picture steganography inside the LAN, while information transmission between a home worker and different gadgets out of the LAN (Internet) will be scrambled. This can provide higher security to the sent confidential information e.g. a picture of a client's face in the LAN network since the presence of touchy data can be concealed utilizing picture steganography. Consequently, the attackers have a smaller amount dubious of the communicated information so as to snoop them (inhaling).

The rest of the chapter is organized as follows. Section 2 includes a few related deals with how the data revelation in IoT gadgets is settled. Section 3 delineates future aspirations. Finally, a conversation and end are attracted in Section 4.

Electronic information transmission may incorporate touchy individual information that might be compromised. Moreover, there are innumerable applications on the web, and numerous locales power clients to obtain pictures that incorporate touchy individual data e.g. portable figures, places, and card data charging. For certain reasons, clients may include individual and safe exchanges e.g. protecting their classified information from developers amid disregarding an open channel, hence requiring positioning and data decency against unapproved section and use. The intermittent methodologies for checking correspondence are cryptography and steganography [2]. Cryptography is the act of utilizing math to encode and unscramble information to keep up posts checked by changing predictable information structure (plaintext) into a muddled (ciphertext) system. The term cryptography originated from the Greek word "cryptos," meaning "wrapped up" and "diagramming hugeness composing." Thus, cryptography's most noteworthy plausible noteworthiness is "covered stating" [3,4]. Any cryptosystem incorporates plaintext, figuring of encryption, computation of unscrambling, the material of code, and a key. The plaintext is a sign or information that is clear (not compacted) in its ordinary structure. Encryption is the course using catches to change over plaintext to discover text.

Figure material outcomes are obtained via encryption by executing the plaintext encryption key. Disentangling is the course to get the plaintext back from the material of the outline, the key is utilized to manage the cryptosystem (setting) information, and is just [3,5] perceived by the sender and beneficiary. While cryptography is exceptionally unbelievable to check information; cryptanalysts may have the option to break the numbers by looking at the amount of component material to return the plaintext [3]. The paper has a clear structure and all aspects are figured going forward. In Section 2, we discuss the substance modules of steganography with respect to the current work. In Section 3, show the III. Similar investigation. Section 5 shows the proposed approach and the results of the investigation. We close the whole paper and present the end in Section 5.

14.2   LITERATURE REVIEW

To determine the data revelation among the IoT gadgets, a few strategies are recommended. The steganography procedure comprises a couple of calculations, changes over simple text (mystery messages) into ciphertext (encryption) at the dispatcher side, and changes it back to simple text (unscrambling) on the beneficiary side [5]. Nevertheless, because of the imperative regular steganography in IoT gadgets, which requests all the more preparation and memory, lightweight cryptography is utilized [6]. Lightweight cryptography is a technique that had some expertise compelled conditions e.g. RFID labels, sensors, contactless brilliant vehicles, and medical services gadgets. In programming usage, inconsequential submissions are favored with little cypher and RAM scope, which does not generally misuse the security-effectiveness compromises. Also, lightweight cryptography needs to embrace IoT on the grounds that it has proven effectiveness from start to finish in correspondence and applications to low asset gadgets. The challenge seen by insubstantial steganography is a level of security, as the diminishing encryption adjusts to the key distance [7].

By and large, the frivolous steganography calculations are separated into symmetrical and unbalanced classifications. In symmetrical encoding, the two players essential to grip mutual importance, and the key is utilized for encryption and decoding. A few instances of symmetrical encryption are PRESENT, HIGHT, and SM1/SM3 in the security framework. A lightweight cryptography convention dependent on XOR activity is proposed to forestall information showing snooping in labels and perusers. The method can shield information transmission from any inactive assault. Other than that, the analysts proposed a plan for an encryption node in IoT gadgets dependent on unique mark highlights and CC2530 [8], which is a genuine framework on-chip (SoC) microcontroller, created by Texas Instruments.

A miniature regulator permits information encryption and unscrambling utilizing AES calculations. In any case, the client's unique mark highlights data recognition whitethorn have mistaken with any corporeal transformed, for example, harmed, soft skin. Likewise, it is suggested to utilize network estimations, all the more especially, to achieve key abstraction, and consequently, any main circulation component is not fundamental. The creators likewise proposed to utilize the keys from encryption/decoding in packed detecting hypothesis to permit encryption and pressure to be done at the same time with CS-based encryption. With this documents trivial and pressure to be achieved at the same time with high efficiency. However, the downside of this technique is that if there is any obstruction or vindictive changes in the properties of the remote channel, the remaking mistake is high as the answers would not be comparative. Other than the snoop, situated an exceptionally (inside a large portion of recurrence frequency) the transporter, he would have the option shrewdly choose the substances that permitted admittance information and the 1s not. Creators are developing another procedure to improve CPABE, which is Obliging Cipher Rule Characteristic Founded Encoding. The idea from IoT permits asset-compelled gadgets to designate expensive exponentiation calculation to other confided-in neighboring gadgets, which are known as "partner hubs." Scientists effectively executed validation and security conspire confirmation and important trade-dependent that allows higher safety execution with proficiency. Furthermore, the correspondence between IoT device remote, Computational Insight (CI), is acquainted with one that is more adaptable, easier, with higher calculation speed and largely adaptive to changing conditions in trade through a courier confidant. This button, which was held entirely mysterious until the two meetings, could then be used to trade hidden texts. In this manner, various significant problems arise and are required to cope with dispersing buttons. Open key cryptography tends to these disadvantages so that clients can impart safely over an open channel without concurring upon a mutual key in advance.

Whitfield Diffie and Martin Hellman published a lopsided main cryptosystem in 1976, which, influenced by Ralph Merkle's job on accessible main distribution, revealed an accessible main comprehension method. This main trade approach, which utilizes exponentiation in a restricted area, has been regarded as the main trade of Diffie–Hellman. The primary structure for using public key or two-key cryptography was the Diffie-Hellman Key Trade Convention. Consequently, it is at some point called Asymmetric encryption. This was the main dispersed down-to-ground method to build a mutual mystery important over a verified (although not personal) route of interchanges without using a previously shared mystery. Steganography and cryptography speak to two strategies for guaranteeing security that has been utilized for quite a while now. Like everything else in the data innovation region, the two are in ceaseless research and improvement. Joining these strategies inside a similar framework is a moderately new heading; however, we can locate a few extraordinary works in writing. One such work is exhibited in paper [4]. The creators propose a framework that will improve the least critical piece (LSB) technique, which is likely the most well-known steganographic strategy. The portrayed framework has a private key transmitted between the sender and the recipient and used to remove the shrouded message. Oakley, J in [1] details four conceivable starting evaluation viewpoints of outer, DMZ, interior, and the basic purposes of the essence. Next, every point of view is differentiated by its capacity to evaluate and abuse vulnerabilities in an association. At that point, the four viewpoints will be thought about by their proficiency and way of assault surface investigation. Finally, inconveniences and preferences of every point of view will be illustrated. Note that aggressive safety assessment is a human-directed process involving both trade and skills, as well as important identification and tools of misuse. In that ability, the method is more workmanship than science, but then it is one of the most important tools that can be accessed to proactively check a scheme owing to the use of subsequent findings that could provide proactive protection of defenselessness. Ogie et al. [2] present an investigation to address this concern by exhibiting an increasingly thorough examination of past security occurrences on basic foundation and modern control frameworks, both as far as the scope of alternatives considered in arranging assaults and the number of episodes tested. An aggregate of 242 revealed security occurrences on basic foundation and modern control frameworks are studied and broken down depending on the proposed arrangement plan introduced in the accompanying approach segment. Han et al. [3], with advancement in data security, concluded that steganography has gotten progressive consideration. It has turned into a pattern that an ever-increasing number of foundations offer steganography strategies in cybersecurity training.

In this paper, we propose coordinating steganography into cybersecurity educational programs. Stenography modules and hands-on labs are planned. It covers the standards of steganography, steganographic strategies, and the premise of steganalysis. Three lab activities incorporate steganography execution in HTML, TCP/IP, and computerized picture. Menon et al. [4] demonstrate an audit on various calculations utilized for steganography. As it appears, every strategy utilizes changed strategies e.g. LSB encoding, pseudo irregular encoding systems, other piece addition procedures to install and various calculations e.g. AES, RSA, RC4, Blowfish calculation, and so forth to change over the plaintext into ciphertext. Every technique has its very own points of interest and burdens. So it is hard to decide the best and the most exceedingly awful one.

This chapter likewise looks at some of them from various perspectives, and which furthermore can be useful in deciding an appropriate technique for explicit utilization. It helps in understanding which calculation is superior to another in a particular circumstance. As per Yari et al. [5], these days, the legal picture examination apparatuses and systems' goal is to uncover the hardening techniques and reestablish the firm faith in the unwavering quality of advanced media. This paper examines the difficulties in identifying steganography in PC crime scene investigation. Open source devices were utilized to examine these difficulties. The trial examination centers around utilizing steganography applications that use the same calculations to shroud data inside a picture. The research finding signifies that, if a specific steganography device is utilized to conceal some data inside an image, and, after that, apparatus B, which uses a similar strategy would not have the option to recuperate the inserted picture. Vegh, L. et al. [6] propose a somewhat new methodology in terms of the framework's security and the one that is utilized in this paper is to integrate cryptography with steganography. When utilizing cryptography alone, the message is encoded, its structure is changed and a key is expected to decode it. When that key is found by a pernicious outsider, the data is undermined. With steganography, the message's presence is covered up, yet the structure does not change. When somebody understands there is a concealed message in whatever record was utilized to shroud it, the data is again traded off. Be that as it may, if the two techniques are consolidated, the security level is a lot higher as both steganalysis and cryptanalysis should be performed so as to locate the first information. A solid security level as portrayed above is the thing that frameworks e.g. CPS need because of their basic application regions [3].

14.3   COMPARISON STUDY

Essentially, providing secret letters is the motive behind cryptography and steganography. Steganography, however, is not similar to cryptography. Cryptography overshadows the content of a noxious person's mystery text, while steganography even includes the message's existence. Steganography should not be confused for cryptography, where we alter the signal to render it obscure to a vindictive person who catches it. The significance of violating the structure is, therefore, diverse in cryptography, when the aggressor can peruse the secret signal, the structure is breached. Breaking a stenographic structure requires the assailant to define the use of steganography and the implanted text to be perused [6]. The composition of a text is blended in cryptography to create it for nothing and to be misleading except when the button to decode is available. It does not attempt to hide or disguise the hidden signal. Cryptography essentially provides the ability to transmit information between individuals in a way that prevents an alien from knowing it. Cryptography can also verify that an individual or thing's character is confirmed. Steganography, on the other hand, does not change the structure of the message of mystery, yet it envelops it within a cover image so that it cannot be seen. For example, a message in ciphertext may raise doubts about the beneficiary, while a message made with stenographic strategies will not be "undetectable." Steganography, in other words, prevents an unintended recipient from suspecting the data occurence. The safety of the steganography system developed also relies on the data encoding framework's mystery [4]. The steganography system is defeated when the encoding system is recognized. Consolidating the approaches is conceivable by storing text using cryptography and hiding the deleted signal using steganography afterward. Without discovering that mystery information is being traded, the subsequent stego image can be transferred. Moreover, irrespective of whether an aggressor would overcome the stenographic approach and recognize the signal from the stego object, it would involve the cryptographic unraveling button to translate the written message in any event. Authors in [1] demonstrate that the two advancements have counter favorable circumstances and disservices. Shockingly, most employments of steganography and research around the point of steganography revolve around ill-conceived purposes. The three greatest zones of ill-conceived steganography revolve around oppression, erotic entertainment, and information robbery. Amid the exploration for this site, the ill-conceived employments of steganography were likewise observed to be on a worldwide scale, included national security, or were done on a scholarly premise so as to all the more likely comprehend the potential risk of steganography whenever made by people with sick aims.

14.4   PROPOSED METHODOLOGY

In a genuine situation, the IP photographic camera gives this ability to verify client expression to open entryways (e.g. worker room entryway, house front entryway). For this situation, the IP photographic camera can acquire face pictures of the clients and direct them to employees for validation using different gadgets outside LAN organization (e.g. distributed storage) for capacity as given in Figure 14.1.

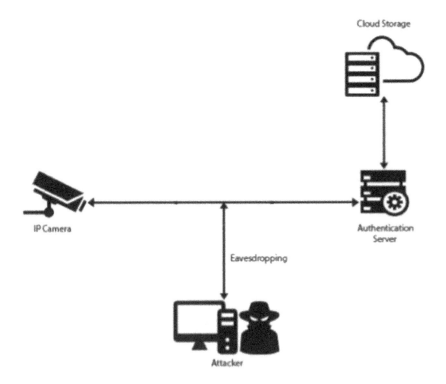

FIGURE 14.1 Actual arrangement.

Any listening in security, particularly in the network privacy data, is communicated. For instance, the achieved picture is as touchy data in Figure 14.1 and any programmer that can effectively get it done has the option to confirm with the verification worker.

Figure 14.2 shows the projected conspire so as make sure about communicated pictures touchy information utilizing IoT gadget. For this situation, picture steganography is utilized so as to make sure about delicate data e.g. client face picture, organization. Furthermore, a home-based worker utilized as an incorporated gadget inside the LAN organization to get the face picture, which is now concealed utilizing picture steganography to scramble all together with them to the gadgets that are situated on the Net (distributed).

As referenced previously, the IoT gadgets cannot handle solid and multifaceted encoding plans, subsequently, frivolous encoding coding techniques utilized. In spite of the fact that the touchy data is in an incoherent organization, aggressors may, in any case, have the option to reveal or decode it with adequate time or handling power. What is more, the encryption does not conceal the presence of data or messages from seeing the assailants. Therefore, the cryptography method utilized an elective safety instrument for moving the information in a safe way. As need be, since the IP camera for the most part manages the picture in our plan, henceforth, we acquaint picture steganography with secure any touchy picture (expression), which needs communicated worker. So as to type the plan basic,

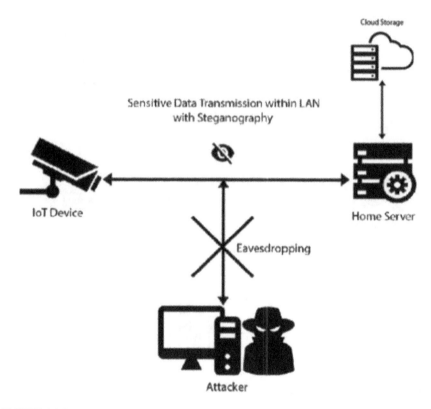

FIGURE 14.2 Future arrangement.

think about Bod, who has the proprietor house, moves toward the IP photographic camera for face identification. After IP photographic camera recognized Bod's face effectively, it supplies it is a picture haphazardly caught with no distinction, it is called a spread picture. Rather than transfer the look picture legitimately to the home-based worker, the look picture covered up (inserted) and chose arbitrary picture (spread picture) with cryptography method, creating an additional picture (stage picture), which covers first look picture. Further, the stage picture is shipped off the home-based worker so as to recover the first expression picture from the stage picture utilizing a similar steganography procedure, yet in an opposite way. Then again, a busybody, Eve, has effectively assaulted Bod's organization capture information broadcast amid IP photographic camera and home worker for any delicate data. She realized that the IP camera capacities has a face discovery authenticator for Bob's home in the front entryway. Accordingly, catches pictures, including the steno picture, yet he does not speculate with that picture as it seems, by all accounts, to be like other pictures. Cryptography is a technique to brand private data and mails imperceptible keep programmers after recognizing. With steganography, assailants wooul not know about the data being communicated through a channel. Furthermore, a few analysts have utilized picture steganography on low-handling power gadgets e.g. inserted gadgets and cell headphones for concealing information. In the accompanying, the least no-teworthy piece (LBS) procedure is depicted as one of the picture steganography strategies for the proposed conspire. It is because of its fewer intricacy contrast with convoluted cryptography techniques just as high limit so as to move more information. The least noteworthy piece (LBS) is a kind of substitution strategy that is one of the most well-known picture of steganography strategies. Actually, the spread picture pixel esteem (e.g. 100) can be spoken to as a line of zeroes and ones (bits). Correspondingly, the touchy data (picture face) can appear as the other series of pieces (1, 0) so as to be supplanted with a portion of the pieces of the spread picture. Indeed, this substitution can happen on the least huge pieces of spread picture pixels, which cannot create a critical change in the presence of the picture (up to 4 LSBs). Conversely, any alteration on the most noteworthy piece (MSB) of a pixel can promptly make a significant corruption on the nature of a picture, which can be distinguished through human discernment. Nonetheless, utilizing the LSBs of pixel to shroud the delicate data can be perceived utilizing measurable examination e.g. x2 investigation as opposed to human vision. This is because of the way that choosing the LSBs for concealing the data is not viewed as randomized, which can produce a marker for the assailant to utilize factual investigation on the picture to discover the modification. To take care of the above issue, rearranged LSB picture steganography is proposed to be used in this plan. With this strategy, the delicate data has less opportunity to be identified by ag-gressors due to the utilizing of spot reversal that prompts improve the nature of stego picture. The creators have proposed two plans for the LBS reversal strategy, which conspire 2 required the spread picture to be gotten by the beneficiary earlier. Since the IP camera is utilized to send the face (delicate data) of the client just a single time, thus, the main plan is adjusted as the picture steganography procedure in this plan. The clarification of the referenced procedure with a model is given as follows. A worker is utilized. In this way, a solitary home worker is adequate in the proposed plan and it would be more cost-effective. Moreover, utilizing a brought together worker is recommended in [31], where the remote hubs are utilized to encode the information while the concentrated worker is utilized to decode the information as it were.

Data Hiding is firmly identified with software engineering, correspondence hypothesis, PC illustrations and picture handling, coding, signal preparing, sci-entific measurements, various media recognition properties, and different fields of information and innovation. As it traverses different subjects, the substance shrouded in this class is of a wide range. The necessities on the course essentials and the premise of information are higher. Not the same as advanced water-marking and encryption innovation, the most noticeable component of steno-graphy is that the private data is implanted into bearers, yet not pulled into the consideration of the others. With respect to the steady struggle of steganalysis, new stenographic advancements are rising. In the meantime, new strategies are highly in demand. Notwithstanding educating the steganography hypothesis, it is likewise important to structure certain exploratory and work on preparing for understudies so as to ace the framework hypothesis and innovation. Steganography should not be confused for cryptography that involves altering the signal to cloud its meaning to vindictive people that prevent it. The significance of violating the structure is distinctive in this particular scenario. In cryptography, when the assailant can peruse the signal of mystery, the structure is breached. Breaking a stenographic structure requires the assailant to acknowledge the use of steganography and the implanted text to be perused. Steganography, as indicated, provides methods for mystery correspondence that cannot be evacuated without substantially altering the data it is entered into. Similarly, the safety of the defined structure for steganography relies on the data encoding framework's mystery. The steganography system is defeated when the encryption system is recognized. Nonetheless, using cryptography and steganography together to include multiple levels of safety is reliably a good method. By entering, a company should be able to encrypt the data and then mount the point signal with the help of the stego button in noise or some other medium. The combination of these two methods will enhance the safety of the embedded data. For instance, restriction, safety, and authority for safe data communication over an accessible circuit will be met by this united study. The amount below delineates the cryptography and steganography mix. The methods to steganography can be split into three kinds: 1) Pure Steganography: It is a scheme that only utilizes the strategy of steganography without consolidating various approaches. It takes a snap to dissimulate information within the distributed holder. 2) Secret Key Steganography: The mixture of mystery important cryptography and steganography strategy is used. This kind of option is used to scramble the mystery signal through the mystery key system and hide the hidden data within the distributed carrier afterward. 3) Community Key Steganography: It is a mixture of the strategy to public important cryptography and steganography. This kind of option is to encode the mystery data using the accessible important methodology and then cover up the stored data within the distributed holder. The Difference between Cryptography and Steganography [9]: Cryptography prevents unapproved parties from discovering the contents of mail, but steganography anticipates revelation of the existence of letters (i.e. cryptography babbles data and realizes that the text passes while steganography generally hides the proximity of hidden data and obscures the signal moving through). Cryptography shifts the mystery text framework while the mystery text system is not adjusted by steganography. Cryptography is a more characteristic development than technology in steganography. Cryptography's most calculations are excellent, yet steganography's calculations are still being generated through particular settings. The solid calculation in cryptography depends on the key size; the greater the key size, the more expensive the processing force is required to decode ciphertext. In steganography, the signal turns out to be recognized when the hidden text is recognized. Cryptography can provide all safety objectives by updating individuals with hash capabilities or verification codes or sophisticated labels in particular and personal key(s). Steganography cannot provide a big part of the safety objectives (integrity, validity, no revocation) autonomous of anyone else without using cryptographic systems. Anyway, it provides autonomous ranking from anyone else on the basis that most of the person concerned understands that the text is wrapped up in what kind of form.

The mystery key steganography method is used in this document to enhance safety by using modified AES and method in [1] that includes PVD, MPK, and MSLDIP-MPK approaches to encode and hide the signal in the distributed image. In this way, if an aggressor asks about the stego picture and attempts to identify the message from the stego picture, the encoded message would in any case require the way to unravel. Proposed Hybrid Algorithm Input: Input the secret information in the format of the message (SI), to define the Cipher Key (CK). Output: To represent the message in the format of Cipher Communication (CC). Phase 1: Step 1. Create CK, this is created by combination of two lists. Step 2. Partition SI to slabs (S1, S2, S3 .... Sn), each and every slab has information about 16 bytes. Step 3. To perform the operation for each Si block do. Step 4. Change every byte to MP(CK) digits (two digits for every byte). Step 5. to segment Si into two state arrays (4*4). Step 6. Applying the filter at each and every state. Stage 7. Make pre-round AddRoundKey, which is a humble bitwiseXOR of the current two states through two subkeys. Step 8. Rehash Step 9. Apply the four changes (SubBytes, ShiftRows, MixColumns, and AddRoundKey) in two states. Stage 10. To play out the nine-round. Stage 11. At last, round actualizes SubBytes, ShiftRows, andAddRoundKey yet MixColumns is erased. Stage 12. Return the digits 9 and 8 in their place in each state. Stage 13. Blend two states to be one square. Stage 14. Convert square to characters by utilizing MPK deciphering (i.e. two digits speak to character). The outcome speaks to cipher-block. Step 15. End Step 16. Link the right now figure block with the past code squares to gather CC. Second Phase: 2 Input: Secret Message M, Cipher Key K, Cover Image C. Yield: Stego Image S. Stage: Phase 1. M has been encoded by utilizing the AES_MPK that takes M and K at that point produces figure text. Stage 2. The code text has been covered up in C by utilizing the strategy in [1] that is consolidating PVD-MPK technique with MSLDIP-MPK strategy and afterward delivers S Experimental Environment: The pre-owned PC with Windows 10 and furnished with a Genuine Intel(R) Core(TM) i5–4210U CPU 1.70 GHz 240 GHz with 8 GB RAM memory. MATLAB R2015b and MATLAB code are utilized to execute the calculation Benchmarks: Several investigations with size 512*512 and 256*256 standard dark scale pictures (Cameraman, Lena, Peppers, Lake, Airplane, and Baboon) were utilized to install a book encoded message.

The message is initially scrambled by AESMPK calculation, and afterward it is covered up by PVD-MSLDIPMPK calculation to be sent. At the recipient, the concealed message is extricated and afterward unscrambled. Benchmarks: Several tests with sizes 512*512 and 256*256 typical dark scale pictures (Cameraman, Lena, Peppers, Lake, Airplane and Baboon) were utilized to encode an sms signal. To begin with, the content is scrambled by the AES-MPK calculation, and afterward the content is disguised by the calculation PVD-MSLDIP-MPK. The covered sign is acquired from the transmitter and afterward unscrambled. Evaluation Parameters: A gauge of nuance (Stego-picture worth) and payload (disguising limitation) is utilized to survey the show of the consolidated computation. Nuance (Stego-picture quality) gauges how regularly contrast (bending) was achieved by data that was put away in the primary spread, where the higher the nature of stego image, the more subtle the message is. The stego-picture exactness could be chosen by utilizing the circumstance (2) characterized by Peak Signal to Noise Ratio (PSNR). On the remote possibility that dim scale picture PSNR is bigger than 36 dB, the human visual structure (HVS) can not recognize the dispersed picture and the stego picture at that stage. Payload (Hiding Capacity) shows how much information can be topped off inside a disseminated picture without unmistakably reshaping the estimation of the scatter picture. Realize that it does not infer that a figuring hides colossal measures of information and produces huge bending in the presentation of the picture. Along these rows, it can be said that a stenographic operation is an extension in the case that it illustrates the development of the payload while maintaining a satisfying verbal character of the stego-picture or improving the performance of the stego-picture at the appropriate, concealing threshold or off possibility that both can be improved [10]. We actualized the open key steganography dependent on coordinating technique in various chosen areas of a picture to demonstrate the presentation of the proposed strategy. In our execution, we utilized 600 × 400 bitmap picture record to conceal 5 KB content information.

As talked about before, both of the two correspondence gatherings should locate the mystery key (stego-key) first by applying Diffie-Hellman open key trade convention to perform abnormal state of security. As in, the 8 bits information will be covered up inside 1 pixel, subsequently the 600 × 400, 24-bit picture record can acknowledge roughly 240,000 bytes of information. This is contrasted and surely understood stego strategy e.g. LSBs which needs 3 pixels to conceal 1 byte of information. We can likewise alter the bit-rate at which we can shroud the information in the chose district. All things considered, the proposed stenographic convention is more effective than LSBs, since the calculation utilized the coordinating strategy to get indistinguishable pixel's bytes. Be that as it may, the proposed strategy resorts to the LSBs technique to appropriate the mystery information on the off chance that if the 8 bit of information is not coordinated with any of the past three bytes (red, green, and blue).

14.5   CONCLUSION

The Internet of Things (IoT) is a standard in the twenty-first century. It is receiving more significance and is being utilized as a component of everyday life. One of the significant worries that is essential to investigate is data classification or security. What is more, programmers can assault an organization because of the presence of weakness inside IoT and little preparation control gadgets can compromise the security of the clients. In this chapter, a plan proposed dependent picture cryptography. Meanwhile, the IP photographic camera with low handling memory capacities utilized as an IoT gadget settles security issues during the transmission between a brilliant gadget and home worker. Because of constraints of brilliant gadgets, particularly memory computational force, the least critical cycle strategy is adjusted. With this strategy, the least critical piece does not bring about significant debasement value with human discernment, just measurable investigation. Also, the likelihood of having a doubt on sent information by the aggressor is utilizing cryptography contrast with the frivolous encoding meanwhile the arrangement information is in coding. Also, tall measure information because the use of picture just as upset LSB picture steganography strategy, which requires fewer pieces for implanting. Since the LBSs choice is in this strategy, the safety of the cryptography calculation can be enhanced. Notwithstanding, choosing the correct bay picture to the picture is one of the downsides of the conspire. As it were, the size of the spread picture and its substance (pixels) must be large enough so as to install the face picture inside itself. Additionally, since the safe correspondence utilizing picture steganography is a single direction component, for a situation that the worker is needed to send any data (e.g. checking the validation) to an IoT gadget (IP photographic camera), data determination is sent as a simple text, which can be easily caught by the attacker. Examination ought to completely execute the frivolous coding related to steganography strategies (double steganography) to give higher security on the communicated information utilizing IoT devices over the group. What is more, the expression picture (creation its scope littler the spread picture) be valuable chosen picture by IP photographic camera as the spread picture. Another safe correspondence model was introduced in this chapter that consolidates methods of cryptography and steganography to give two layers of security so that the steganalyst cannot achieve plaintext without knowing the mystery button to decode the ciphertext. Initially, the mystery data was recorded using the AES MPK, then, using hybrid methods to cover up the deleted data in the bleak image. Because of this mixture, the data of the mystery can be transmitted over the open channel in view of the reality that the ciphertext does not appear aimless but rather disguises its nature by using steganography to conceal ciphertext in the images. Test findings showed that our suggested model can be used to shroud considerably more data than other existing methods and that the graphic character of the stego image is also enhanced, although it is strong for the communication of mystery information. We anticipate adding the suggested method to noise and television in a subsequent job. In addition, we expect the suggested method to be upgraded to render the threshold lower than it is, while maintaining the PSNR equal or greater.

REFERENCES

1. Oakley, J. (2018). Improving hostile digital protection appraisals utilizing shifted and novel instatement viewpoints. Proceedings of the ACMSE 2018 Conference on - ACMSE '18. doi:10.1145/3190645.3190673.
2. Ogie, R. I. (2017). Cyber Security Incidents on Critical Infrastructure and Industrial Networks. Proceedings of the 9th International Conference on Computer and Automation Engineering – ICCAE '17. doi:10.1145/3057039.3057076.
3. Han, D., Yang, J., & Summers, W. (2017). Inject Stenography into Cybersecurity Education. 2017 31st International Conference on Advanced Information Networking and Applications Workshops (WAINA). doi:10.1109/waina.2017.30.
4. Menon, N., & Vaithiyanathan. (2017). A survey on image steganography. 2017 International Conference on Technological Advancements in Power and Energy (TAP Energy). doi:10.1109/tapenergy.2017.8397274.
5. Yari, I. A., & Zargari, S. (2017). An overview and computer forensic challenges in image steganography. 2017 IEEE International Conference on Internet of Things (iThings) and IEEE Green Computing and Communications (GreenCom) and IEEE Cyber, Physical and Social Computing (CPSCom) and IEEE Smart Data (SmartData). doi:10.1109/ithingsgreencom-cpscom-smartdata.2017.60

6. Vegh, L., & Miclea, L. (2014). Securing communication in cyber-physical systems using steganography and cryptography. 2014 10th International Conference on Communications (COMM). doi: 10.1109/iccomm.2014.6866697

7. Salau, A.O., Marriwala, N., & Athaee, M. (2021). Data Security in Wireless Sensor Networks: Attacks and Countermeasures. *Lecture Notes in Networks and Systems*, vol. 140. Springer, Singapore. https://doi.org/10.1007/978-981-15-7130-5_13

8. Du, X., & Chen, H. H. (2008). Security in wireless sensor networks. *IEEE Wireless Communications*, vol. 15, no. 4, pp. 60–66.

9. Yadav, V., Ingale, V., Sapkal, A., & Patil, G. (2014). Cryptographic Steganography. *Computer Science & Information Technology*, pp. 17–23.

10. Sharma, M. H., Mithlesharya, M., & Goyal, D. (2013). Security Image Hiding Algorithm using Cryptography and Steganography. *Journal of Computer Engineering*, vol. 13, no. 5, pp. 1–6.

15 Optimal Automatic Power Generation Using Modified Hybrid Soft Computing Techniques

Dr. Cheshta Jain Khare[1], Dr. Vikas Khare[2], and Dr. H.K. Verma[3]
[1]Assistant professor, SGSITS, Indore
[2]Associate Professor, STME, NMIMS, Indore
[3]Professor, SGSITS, Indore

15.1 INTRODUCTION

Power turned into a subject of logical enthusiasm in the late seventeenth century, crafted by William Gilbert. Throughout the following two centuries, quantities of critical revelations were made, including the radiant light and the unstable heap. Presumably, the best disclosure regarding "Power Engineering" originated from Michael Faraday, who, in 1831, found that a change in attractive transition prompts an EMF in a circle of the wire-a standard, known as electromagnetic enlistment, that clarifies how generators and transformers function. In other words, control building, additionally called control framework designing, is a sub-part of vitality designing and electrical designing that has arrangements with the generation, transmission, distribution, and the use of electric power and electric gadgets associated with such frameworks, including transformers, engines, and generators.

The greater part of the power generation happens at producing stations that may contain in excess of one such alternator-turbine combination. Cutting-edge control frameworks are confounded systems with many producing stations and load focuses being interconnected through power transmission lines. One of the primary challenges in the electrical power framework is that the measure of dynamic power is expended, in addition to losses it should constantly rise to the dynamic influence delivered. In the event that more power would be created than expended, the frequency would rise and the other way around. Indeed, even little deviations from the ostensible frequency esteem would harm the synchronous machines and different apparatuses. To compensate for such type of problem, load frequency control and the automatic generation control (AGC) concept are being used in recent trends. In an electric power framework, AGC is a framework for modifying the power yield of numerous generators at various power plants, in the response to change in loads.

DOI: 10.1201/9781003140443-15

The adjusting can be balanced by estimating the system frequency, in the event that it is expanding, more power is being produced than utilized, which makes the whole machine in the framework quicker. Prior to the utilization of AGC, one producing unit in a framework would be planned as the directing unit and would be physically changed in accordance with control the harmony among generator and load to keep up the framework to the frequency at the desired value. Where the grid has attached interconnection to the adjoining control areas, AGC keeps up the power trades over the tie-lines at the planned levels. An AGC system can take into account computer-based control systems and diverse inputs, such as the most economical units to change, the synchronization of forms of thermal, hydroelectric, and other genera-tions, and also limitations relevant to the system's constancy and the capacity to interconnect to other power grids. To enhance the tie-line bias, the AGC signal is formed by the feedback of an integrated weighted sum of inter-area power flow and frequency error. This error is called an area control error (ACE) [1]. To improve the performance of such a system, the use of various apparatuses with large capacity has increased extensively. Nonlinear characteristics of such loads may create a serious problem when they are connected to a power system [2].

15.2 MODELING OF LOAD FREQUENCY CONTROL SYSTEM WITH AND WITHOUT THE EFFECT OF DEAD BAND

The fundamental goals of AGC are:

    i. To retain the frequency within specified limits
    ii. To manage tie-line power flows
    iii. To allocate the loads among the generating units of a multi-area system

In AGC, the change in frequency (Δf) is amplified and transferred to manage the valve position of the turbine or prime mover. The changes in the output power of the turbine establish the real power balance, as shown in Figure 15.1.

When the system frequency deviates from its scheduled value, due to a sudden change in load, then an error signal is created, which is used to alter the valve position. The block diagram representation of the single area control system is shown in Figure 15.2, along with the speed control mechanism [3]. Further, Figure 15.3 represents the block diagram of the single control area along with the additive supplementary control loop.

These figures clearly illustrate the change in governor output (ΔP_g) with respect to changes in reference power setting (ΔP_{ref}), which can be represented by the following equation:

$$\Delta P_g = \Delta P_{ref} - \frac{1}{R}\Delta f \ \text{MW} \tag{15.1}$$

The constant R (Hz/MW) is referred to as regulation or droop. The droop is the percentage change in frequency for a 100% improvement in the turbine generator

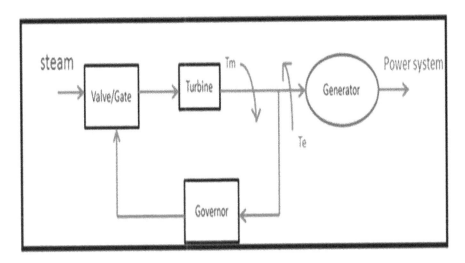

FIGURE 15.1 The schematic representation of AGC.

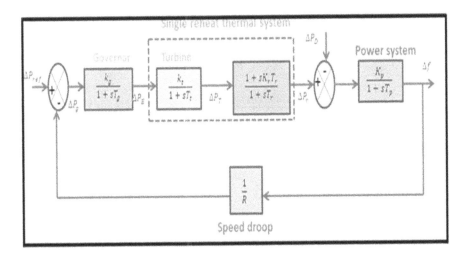

FIGURE 15.2 The block diagram representation of single area AGC.

set's power output. A standard drop configuration for a turbine generator set is 4%, which means that a 4% rise in a turbine's rotational speed would shift the controller value from the fully closed position to the fully open position or if the frequency decreases by 0.3% due to sudden load change, then the power output of turbine generator set would be increased by 7.5% $\left(0.3 \times \frac{100}{4}\%\right)$.

The output of the governor valve (ΔP_E) can be represented using Figure 2.3 as [3]:

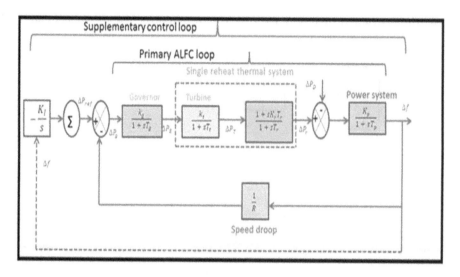

FIGURE 15.3 The block diagram representation of single area AGC with the supplementary control loop.

$$\Delta P_E = \frac{k_g}{1 + sT_g}\Delta P_g \tag{15.2}$$

Where k_g is gain and T_g is the time constant of the governor valve. Turbine power maintains air gap power to control speed or frequency. The change in turbine power depends upon changes in valve power, because of this dynamic, the turbine can be expressed by transfer function as [4]:

$$\frac{\Delta P_T}{\Delta P_E} = \frac{k_t}{1 + sT_t} \tag{15.3}$$

In a reheat thermal system output, power is expressed in terms of the transfer function as [5]:

$$\frac{\Delta P_r}{\Delta P_T} = \frac{1 + sK_r T_r}{1 + sT_r} \tag{15.4}$$

The output of generators and electric load constitutes the power system. The load on the system consists of a frequency-dependent and an independent component. The change in load can be written as $\Delta P_L = \Delta P_D + \Delta P_f$, where ΔP_D is frequency-independent component of load and ΔP_f is a frequency-dependent component of the load. Hence, the characteristics of the load (also called damping constant) can be expressed as a percentage change in load for percentage change in frequency as [3]:

$$D = \frac{\partial P_f}{\partial f} \text{ MW/Hz} \tag{15.5}$$

The change in load demand (ΔP_D) causes a change in generator output to match the required demand. As a result, the frequency will change. Since the kinetic energy is proportional to the square of the speed, the equal-area criterion requires that the turbine power will be increased. This speed governing system can be modeled by the swing equation [3]. The difficulties associated with frequency control in an interconnected area are more imperative than a single or isolated area. Nowadays, all power systems are tied to adjacent areas, and the problem of load frequency control becomes a mutual responsibility. In an interconnected system, each area should carry its own load under normal operating conditions, and when sudden load changes, then the required energy is being borrowed from the kinetic energy of the mutually connected areas. Such interconnection is achieved via tie-lines. In general, there are a number of areas that may be interconnected through tie-lines and, therefore, control of tie-line power flow with control of frequency changes is needed. The tie-line power flow between j^{th} area and k^{th} area can be expressed by:

$$P_{tie,jk} = \frac{|V_j||V_k|}{X} \sin(\delta_j - \delta_k) \tag{15.6}$$

Where δ_j and δ_k are the angles of end voltages V_j and V_k, respectively. For a small change in the angles, the tie-line power flow changes as:

$$\Delta P_{tie,j} = T_j(\Delta \delta_j - \Delta \delta_k) \tag{15.7}$$

Where, T is electric stiffness of synchronous machines which is defined as [3]:

$$T_j = \frac{|V_j||V_k|}{X} \cos(\delta_j - \delta_k) \tag{15.8}$$

The frequency deviation Δf relates with change in angle ($\Delta \delta$) by following equation:

$$\Delta \delta = 2\pi \int \Delta f \, dt \tag{15.9}$$

Hence, the change in tie-line power is expressed in Laplace transform using equations (15.7 and 15.9):

$$\Delta P_{tie,j}(s) = \sum_{k=1}^{n} \frac{2\pi T_{jk}}{s}(\Delta f_j - \Delta f_k) \tag{15.10}$$

Where n is the number of interconnected areas. Since the error in the tie-line power flow is the integral of the frequency difference between j^{th} and k^{th} areas, therefore to control frequency error at a predetermined value, any steady state errors in the system frequency would outcome in tie-line power errors. Hence, in the sequence of the tie-line power perturbation must be included in control input. Consequently, an area control error (ACE) of the jth area is defined as [6]:

$$ACE_j = \Delta P_{tie,j} + B_j \Delta f_j \qquad (15.11)$$

Where B_j is the frequency bias or frequency characteristic of a jth area which is based on the area's natural regulation characteristics $((1/R)+D)$ corresponding to the forecasted peak load of the coming year in each area [6]. The average frequency bias setting employed is about 2% per 0.1 Hz based on the expected peak load and spinning reserve. In the case of a multi-area system, each area is set for a different load setting and if the load is changed in any area which is reflected on the other area through tie-lines because of this dynamic response of frequency is changed. This change in frequency is reset at a predetermined value through some controller with an optimum value of frequency bias setting. A linear model is enough for dynamic illustration of the system under normal operation. Figure 15.4 shows the block diagram for j^{th} area representation of the reheat thermal system.

Effect of governor dead band: In electric power, system generation can be changed at a maximum rate. Therefore, the dynamic performance of the speed governor system is affected by the inclusion of the dead band. The governor dead band is a function of time when there is no corresponding change in valve position in the overall magnitude of a continuous change in speed in the band. For more practical evaluation, the effect of the governor dead band has to be added, which makes the system nonlinear. Dead band AGC is represented in a percentage of the rated turbine speed. The dead band is caused by Coulomb friction and backlash effect in various governor linkages and by valve overlap in a hydraulic relay. The consequence of the dead band on the speed governor response depends on the

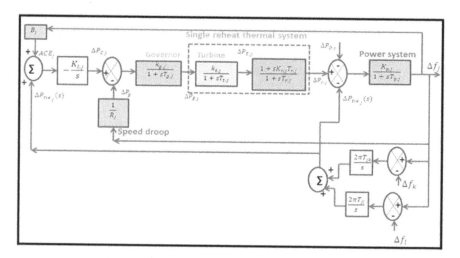

FIGURE 15.4 Complete block diagram of single area reheat thermal AGC of jth area.

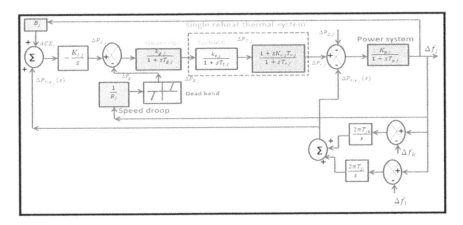

FIGURE 15.5 Block diagram of jth area AGC imply dead band in speed governing loop.

magnitude of the change in frequency. If a change in frequency is much smaller, it will remain in the band; hence, speed control will be inactive. For a small change in load, the setting causes a random response in each generating unit. Therefore, the inclusion of speed governor dead bands results in random frequency deviations [6]. The block diagram of jth area with the dead band is depicted in Figure 15.5. The change in frequency Δf_j changes gradually with time so that amid a specific time interim each area operates completely within the dead band or outside the band. Subsequently, the system turns into principally a linear system when the governor dead band is considered.

15.2.1 PROBLEM FORMULATION

In the present workload, frequency dynamics are illustrated by developing mathematical models of the interconnected reheat thermal system in the state variable form. Since both linear and nonlinear systems are studied in the proposed research work, therefore, the state space model is suitable for problem formulation. The tie-line power perturbation can be tacit as an added interruption to any j^{th} area. The area-wise state variable is defined as follows [7]:

$$X_1 = \Delta f_j, \ X_2 = \Delta P_{E,j}, \ X_3 = \Delta P_{T,j}, \ X_4 = \dot{X}_3, \ X_5 = \Delta P_{tie,j}, \ X_6 = \Delta P_{C,j}.$$

The state space equation of jth area can be written in the following form:

$$[\dot{X}]_j = [A]_j [X]_j + [U]_j \tag{15.12}$$

Where, $[X]_j^T = [X_1, X_2, X_3, X_4, X_5, X_6]_j$ and $[A]_j$ is state matrix of jth area which can be write using state equation of Figure 2.6 as follows [3]:

$$[A]_j = \begin{bmatrix} \frac{-1}{T_{p,j}} & 0 & \frac{K_{p,j}}{T_{p,j}} & 0 & -\frac{K_{p,j}}{T_{p,j}} & 0 \\ a_{21} & \frac{-1}{T_{g,j}} & 0 & 0 & 0 & \frac{1}{T_{g,j}} \\ 0 & 0 & 0 & 1 & 0 & 0 \\ a_{41} & \left(\frac{1}{T_{t,j}T_{r,j}} - \frac{K_{r,j}}{T_{g,j}T_{t,j}}\right) & \frac{-1}{T_{r,j}T_{t,j}} & -\left(\frac{T_{r,j}+T_{t,j}}{T_{r,j}T_{t,j}}\right) & 0 & \frac{K_{r,j}}{T_{g,j}T_{t,j}} \\ \sum_{\substack{k=1\\k\neq j}}^{n} 2\pi T_{jk} & 0 & 0 & 0 & 0 & 0 \\ -K_{I,j}B_j & 0 & 0 & 0 & -K_{I,j} & 0 \end{bmatrix} \quad (15.13)$$

And matrix $[U]_j$ is defined as follows:

$$[U]_j = \begin{bmatrix} -\frac{K_{p,j}}{T_{p,j}}\Delta P_{D,j} & U_{2,j} & 0 & U_{4,j} & -\sum_{\substack{k=1\\k\neq j}}^{n} 2\pi T_{jk}\Delta f_k & 0 \end{bmatrix} \quad (15.14)$$

Where the value of a_{21}, a_{41}, U_2 and U_4 depends whether the effect of dead band is consider or not.

2 For the linear system (the dead band not considered) these values are expressed as [7]:

$$a_{21} = -\frac{1}{R_j T_{g,j}} \quad (15.15)$$

$$a_{41} = -\frac{K_{r,j}}{R_j T_{g,j} T_{t,j}} \quad (15.16)$$

and U_2, U_4 both are equal to zero.

2 For the nonlinear model (considering dead band) there will be two $[A]$ matrices for each area, one considered operation within the dead band and another considered the operation exterior the dead band region.

i. When operating within the dead band: there is no signal available corresponding to a change in frequency. Hence, values of matrix element a_{21}, a_{41}, U_2 and U_4 become zero.

ii. Operation outside the band: when a deviation in frequency is greater than 0.06% (DB) then the values of elements becomes [8]:

$$a_{21} = -\frac{1}{R_j T_{g,j}}$$

$$a_{41} = -\frac{K_{r,j}}{R_j T_{g,j} T_{t,j}}$$

$$U_2 = \frac{DB}{R_j T_{g,j}} sign(\Delta f_j) \tag{15.17}$$

$$U_4 = \frac{K_{r,j}}{R_j T_{g,j} T_{t,j}} DB sign(\Delta f_j) \tag{15.18}$$

This chapter explained two test systems one has two-area systems and the second have four-area systems with and without the effect of speed governing a dead band of 0.06%.

15.3 OVERVIEW OF SOFT COMPUTING TECHNIQUES

The soft computing method is a promising set of techniques that expect to build up easiness for indistinctness, vulnerability, and fragmentary truth to accomplish strength, tractability, and aggregate minimal effort. Soft computing is essentially an optimization technique to acquire the solution to problems that are very difficult to answer. Optimization is the process of controlling the inputs to find the best possible output. The evaluation question relates to the numerous issues of combinatorial. The approach used in problems of computational geometry can be categorized into two classes, the specific methods first and the approximate (heuristic methods) second. The specific methods established a solution to the problem at hand, but not sufficient for real-life problems; because of their dynamic environment, it takes significant calculation time. Heuristic approaches are ideal for realistic use to find correct solutions under acceptable estimation times [9]. Meta-heuristics, which have worked successfully on big and multifaceted issues in real life, are another form of methodology. The classification of such search methodologies is shown in Figure 15.6

For frequency control in AGC, various control strategies e.g. integral, proportional integral (PI), PID have been developed. Due to a fixed structure and constant parameter, the conventional controller maintains system frequency for one working condition. Since most of the power system components are nonlinear, this controller may not able to provide desired performance for other operating conditions. To eliminate the above problem, a number of decentralized load frequency controller was developed which is optimized by soft computing techniques.

Many researchers have applied the optimal control concept, variable structure control, artificial neural network, fuzzy logic to select the controller parameter for the conventional linear AGC system. Little work has been accounted for to handle the troubles related to a nonlinear model utilizing heuristic transformative procedures. Far from each other from propels in charge ideas, there have been many changes amid the most recent decade e.g. deregulation of the power business, and utilization of superconducting magnetic energy storage, wind turbines, photovoltaic cells as non-traditional energy wellsprings of electrical energy of the framework. Therefore, control strategy associated with an AGC problem has changed.

A general scheme structure of evolutionary algorithms is shown in Figure 15.7

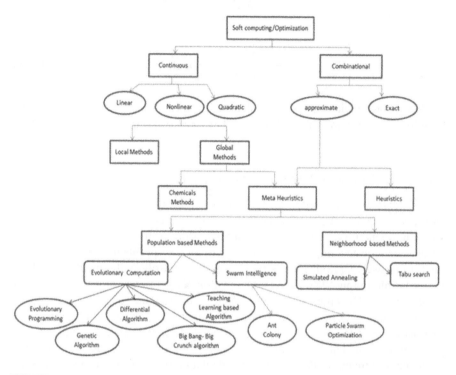

FIGURE 15.6   Classification of search methodologies for optimization [10].

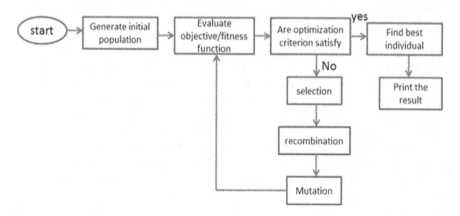

FIGURE 15.7   Scheme structure of evolutionary algorithm [9].

A number of evolutionary methods were explored for many engineering applications in the last few decades, which are briefly mentioned as follows:

Particle swarm optimization (PSO): PSO was first projected by Kennedy and Eberhart [11]. Like transformative calculations, PSO strategies lead a search utilizing a population of particles. The particles in the PSO change their ebb and flow position by flying around in a pursuit space until the greatest emphasis check is not reached. The PSO is propelled by the capability of flocks of birds to accomplish a new goal. PSO is a stochastic search algorithm based on the population that is used to obtain the optimal solution which is built on the behavior of animals' social behavior, such as bird flocking and fish schooling. In contrast to the other evolutionary algorithm, this approach is simple to apply with very limited parameters that need to be modified in consecutive iterations.

There are two levels of a simple PSO, discovery and exploitation. In the discovery process, particles look for the most promising regions and particles travel into the best location in the exploitation phase. By altering their location with regard to the optimum location of particles, PSO explores the global best (gbest) value among the different locations of particles. The personal best value of each particle (pbest) expresses its data through its neighbors to the rest of the particles. The overall best particle location then steadily absorbs the rest of the particles corresponding to the modified velocity of every particle, which depends on the values of global best and pbest. The competency of the algorithm depends on the approach used to choose the parameters for the subsequent iteration. Three steps are needed for the simple PSO algorithm, namely, particle generation, positions, and velocities; second, velocity update; and third, position update [11]. The category of random particle locations is initialized with the PSO (x_i^k) and velocities (v_i^k) as expressed in the following equations [11], between the upper and lower limits of the design variable values

$$x_i^k = x_{min} + rand\,(N, d).\,_*(x_{max} - x_{min})   \tag{15.19}$$

And

$$v_i^k = v_{min} + rand\,(N, d).\,_*(v_{max} - v_{min})   \tag{15.20}$$

Where N: number of population. d: number of parameters to optimize. k: current iteration count. x_{min} and x_{max}: minimum and maximum value of particles in search space. v_{min} and v_{max}: minimum and maximum value of the position of particles to move in search space.

For the next (k+1) iteration, the next step is to change the velocities of all particle positions using the particle fitness values that act as particle positions. This value of the fitness function defines which particle in the current swarm (*iteration*) has the best global value (*gbest^k*) and also determines the best location of each particle (*pbest_i*).

After determining the two best values, the particles change their velocity and positions corresponding to (*i*th) particle using the following equation:

$$v_i^{k+1} = wv_i^k + c_1 r_1 (pbest_i^k - x_i^k) + c_2 r_2 (gbest^k - x_i^k) \qquad (15.21)$$

The two different random values between 0 and 1 are assigned as a constant r1 and r2. C1 and C2 are constants of acceleration which can be set at 2. Such constants help to shift particles' position near the best value available ($gbest^k$) and "w" is the weight of inertia used to balance the best value between previous and present. Weight change of inertia in an efficient iteration as iteration [12,13]:

$$w = w_{max} - \frac{(w_{max} - w_{min})}{itermax} * iter \qquad (15.22)$$

Where the actual number of iterations is itermax. The upper and lower limit of inertia weights denoted as w_{max} and w_{min}, which are set at 0.9 and 0.4, respectively. Particle positions are currently updated using the following equation:

$$x_i^{k+1} = x_i^k + v_i^{k+1} \qquad (15.23)$$

Implementation of PSO algorithm

Step 1: Optimization parameters are initialized, such as population size (N), maximum iteration number (itermax), number of variables (d), search space set (x_{min} and x_{max}), etc.
Step 2: Set $k=1$ and randomly generate particle's position and their velocities using Equations 15.1 and 15.2, respectively.
Step 3: Determine the fitness function corresponding to each population further determine personal and global best value of kth iteration, that is, $pbest_i^k$ and $gbest^k$.
Step 4: Change the velocity of every particle and their positions using Equations 15.3 and 15.5, respectively and set $k=k+1$.
Step 5: Stop the procedure if either max iteration count reached or optimum value corresponding to mean value of fitness function or median value of function value becomes equal to global value otherwise go to step 3.

Chaotic PSO (CPSO):
In recent years, the PSO has become much more popular in different kinds of applications because of its easy implementation. However, the traditional PSO is highly dependent on its parameter and suffers the problem of being trapped in local optima. With multi-dimensional functions when local optima are found in regions where fitness function varies rapidly, the PSO fails at this stage. To solve this problem, the chaotic PSO (CPSO) method has been introduced. General PSO depends on its parameter. In PSO after certain iterations, the parameter sets are

approximately identical and no improvement is noticed [152]. Optimization based on chaotic sequence can be a better way to provide diversity in the population. A chaotic sequence for inertia weight and constriction factor for the optimization of gains is used.

i. *Adaptive inertia weight factor (AIWF):*

To achieve improved variety, CPSO uses chaotic sequences for inertia weight and constriction factor [152]. Inertia weight "w" is dynamically modified to monitor *gbest* value according to the best fitness value utilizing the following expression:

$$w_i^k = w_{min} + \frac{f_{pbest}^k |f_i^k - f_{pbest}^k|}{f_i^k |f_i^k - f_{gbest}^k|} \quad (15.24)$$

Where w_i^k is inertia weight of ith population at k^{th} iteration. w_{min} is minimum inertia weight. f_{pbest}^k is fitness function corresponding to pbest values at k^{th} iteration. f_i^k fitness function of ith population at k^{th} iteration. f_{gbest}^k fitness functions of gbest value at k^{th} iteration.

ii. *Adaptive Constriction Factors:*

A fitness function according to *pbest* and *gbest* values at k^{th} iteration is heavily dependent on the constriction factor c1 and c2. Such considerations are revised as [14,15]:

$$c_{1i}^k = sqrt\left(\frac{f_i^k}{f_{pbest}^k}\right) \quad (15.25)$$

$$c_{2i}^k = sqrt\left(\frac{f_i^k}{f_{gbest}^k}\right) \quad (15.26)$$

So, velocity up gradation modified as:

$$v_i^{k+1} = w_i^k v_i^k + c_{1i}^k r_1(pbest_i^k - x_i^k) + c_{2i}^k r_2(gbest_i^k - x_i^k) \quad (15.27)$$

The algorithm of CPSO is as follows:

Step 1: Set the population size and number of iteration, etc.
Step 2: Randomly generate "N" particles ($x_i(0)$, $i = 1, 2 ...,N$) with uniform probability over the search space of optimized parameter [x_{min}, x_{max}]; similarly, initialize the velocities of all particles as given in Equation 15.2:
Step 3: Evaluate the fitness function of each particle at k^{th} iteration.
Step 4: Determine the *gbest* value and *pbest* value.
Step 5: Find the fitness function corresponding to best values.

Step 6: Evaluate AIWF and constriction factors using Equations 15.6 to 15.8.
Step 7: Change the velocity of each particle by Equation 15.9.
Step 8: Each particle changes its location in compliance with equation, based on modified speeds (Equation 15.5). If a particle in some dimension exceeds the position limit, set its position at the correct limit.
Step 9: If the last improvement in the best solution is larger than the pre-specified number, or if the maximum number of iterations is reached, stop the method, or proceed to step 4.

Differential evaluation: Differential evaluation (DE) was exhibited as a heuristic enhancement technique that has been utilized to limit nonlinear or non-differentiable capacities [16]. The goal of DE is to discover optimum estimation of parameters by upgrading the execution of a system, under some given conditions. In DE, the distinction of arbitrarily chose a pair of object vectors is utilized to give a transformation while different heuristics calculations utilize probability distribution factor. The critical preferences of DE are its ability to seek successful worldwide optimal points, without arranging and effortlessly illuminate non-differentiable time-dependent noisy fitness function [17].

DE has been described as a technique of heuristic optimization used to minimize nonlinear or non-differentiable functions [16]. While optimizing the performance of a system, the goal is to identify such a set of system parameter values for which, under certain conditions, the overall performance of the system will be best. It utilizes the randomly sampled pair of object vector differences to provide mutation, while other EA's use functions of a probability distribution. DE's main functions are its efficient global optimization capability, efficient algorithm without sorting, and non-differentiable noisy time-dependent objective function that is easily handled. This operates through a simple cycle of phases as follows:

Stage 1: Initial Population

EquationThe parameters governing the system are presented as $x = [x_1, x_2, ---- x_N]$ in vector form, each x_i parameter being an actual number. In the N-dimensional search space, DE searches for a global optimal point. It initializes the vector parameter randomly within the search space under the uniform distribution of probability limited by the prescribed minimum and maximum bound as given 15.1:

Stage 2: Mutation

The main feature of DE is the manner in which the new population, called a mutation, is generated. The weighted distinction between two random vectors added to the third random vector known as the donor vector obtains a mutant vector $(x_{new,i}^{k+1})$ [17]. Some predetermined test vectors are mixed with these newly generated vectors (v_i^k). Three separate vector parameters x_{r1}, x_{r2}, and x_{r3} are randomly generated to generate donor vector for each ith target from the current (k^{th}) population. The difference is scaled by constant F for any two of these three vectors, and the scaled difference is added to the third one as:

$$x_{new,i}^k = x_{r3}^k + F. (x_{r1}^k - x_{r2}^k)$$

(15.28)

Where $r_1 \neq r_2 \neq r_3 \neq i$ are mutually exclusive. The scale factor F controls the scale of difference $(x_{r1}{}^k - x_{r2}{}^k)$.

Stage 3: Crossover:

After the donor vector is formed by mutation, its components are switched from the current generation to the target vector, which is the parent vector. According to crossover likelihood, a binomial crossover is performed for any N element. There is a binomial distribution of the number of parameters inherited from the donor. By producing new vectors from current object vector parameters, crossover strengthens previous achievements. The likelihood of a crossover (CR) is used to determine whether or not the new vector can be recombined.

$$v_i^k = \begin{cases} x_{new,i}^k & if \quad rand\,[0,\,1] \leq CR \\ x_i^k & otherwiswe \end{cases}$$

(15.29)

Stage 4: Selection or Recombination:

By evaluating the fitness function, the newly generated vector is obtained. Selection determines whether, for the next iteration, the goal vector is sufficient or not. The value of the new generation fitness function $f(v_i^k)$ is compared to the previous fitness function $f(x_i^k)$ of the respective vectors. If the newly generated vector fitness functions have a lower value than the previous one then the former vector is replaced by the newly generated vector as:

$$x_i^{k+1} = \begin{cases} v_i^k & if \quad f(v_i^k) \leq f(x_i^k) \\ x_i^k & if \quad f(v_i^k) > f(x_i^k) \end{cases}$$

(15.30)

For each generation, the right parameters are assessed to monitor the progress of minimization. DE typically uses set population sizes in the process. Population size selection (pop) is very critical. Pop should be as small as possible to produce rapid numerical performance, but such a small size can lead to premature convergence. Therefore, the population scale of 20d in the engineering application produces inherently stronger solutions.

Big Bang–Big Crunch (BB-BC): The Big Bang–Big Crunch advancement is a heuristic transformative calculation based on population. This strategy is created by Osman and Eskin [18,19] which can deal with the multidimensional issue effectively with the fast convergence rate. The idea of an algorithm depends on the arrangement of the Universe Kwon by the Big Bang hypothesis. As indicated by this hypothesis, when the universe was a circle with infinite radius and thickness, then because of some inside forces, the existed mass is exploded greatly, which is called the Big Bang, and billions of particles moved outwards [20]. A gravitational power emerged because of the spreading of particles, which relies upon the masses of two bodies considered and the separation between them. As extension happens the gravitational power on all particles diminishes [21]. In view of disseminated kinetic energy and extension of gravitational energy between particles, the kinetic energy of particles is contracting. At this point, all particles fall into a solitary molecule called the Big Crunch.

BB–BC is a novel algorithm based on the universe's evolutionary theories: the Big Bang hypothesis and the Big Crunch theory. According to these hypotheses, the expansion of the universe ceases if the gravitational force is greater than the energy produced by the Big Bang, which is accompanied by a contraction. The world would be limited to a single point by this.

The optimization of the Big Bang–Big Crunch is an evolutionary process focused on a heuristic population. Osman and Eskin[18] have developed this methodology that can easily handle multidimensional problems with very rapid convergence. This algorithm is based on the Big Bang hypothesis of the origin of the universe. The world was once a sphere with an infinite radius and mass, according to this hypothesis. The current mass is massively exploded, dubbed the Big Bang, due to many internal forces, and billions of particles travel outwards. If particles continue to disperse, a gravitational force occurs, which depends on the masses and distance between the two called bodies. The gravitational force on each particle falls as expansion takes place and the kinetic energy of contraction dissipates rapidly [21]. The kinetic energy resulting in particles shrinking is offset by gravitational energy between particles due to expansion. All particles collapse into a single particle named Big Crunch at this point. In a simple series of steps, this algorithm works as:

Stage 1 (Big Bang phase):

In this method, the initialization is comparable to other evolutionary methods. An initial candidate population is generated randomly over the entire search space, as given by eqn-1.1. The Big Bang process activity is known as energy dissipation. Initialization randomness is the same as energy dissipation in nature, only that this dissipation creates confusion from structured particles and uses this randomness to create new candidate solutions (disorder or chaos). To not skip the global argument, the number of people in the population must be high enough.

Stage 2 (Big Crunch phase):

As a convergence operator, the Big Crunch stage will come. There is only one performance, called the center of mass, at this stage. The center of mass is the weighted average solution of the candidate as a [20]:

$$X_{com} = \frac{\sum_{i=1}^{pop} \frac{x_i^{(k)}}{f_i}}{\sum_{i=1}^{pop} \frac{1}{fi}} \tag{15.31}$$

Where X_{com} is the position of the center of mass. x_i^k is the position of ith candidate in N dimensional search space. f_i is fitness function value of ith candidate. *Pop* is the number of candidate population.

Stage 3 (generate new population):

The Big Bang process is typically split around the center of gravity. By inserting or subtracting a standard random number, the new candidate near the center of mass is determined as:

$$x_i^{new} = \beta. X_{com} + (1 - \beta)x_{best} + \frac{rand. \alpha (x_{i(max)} - x_{i(min)})}{iteration\ step} \tag{15.32}$$

Where α is constant which limiting the size of search space.

β is used to controlling the deviation of the global best solution x_{best} from the new candidate solution. The best solution x_{best} influences the direction of search [20].

Stage 4 (selection or recombination):

Apply the selection criteria now. Selection decides whether, for the next iteration, the current candidate is suitable or not. The value of the current-generation fitness function $f(x_i^{new})$ is compared to the fitness function $f(x_i^{privious})$ of the previous iteration for the entire individual concerned. If the newly generated candidate's fitness functions have a lower value than the previous one then the former candidate is replaced by the newly generated candidate as a:

$$x_i^{next\ iteration} = \begin{cases} x_i^{privious} & if\ f(x_i^{privious}) \le f(x_i^{new}) \\ x_i^{new} & if\ f(x_i^{new}) < f(x_i^{privious}) \end{cases} \tag{15.33}$$

Teaching learning-based optimization: Optimization based on teaching–learning (TLBO) is suggested by Rao et al. [22]. This approach relies on two variables i.e. the willingness of teachers to teach a variety of students to communicate with each other. There are two phases of the TLBO (i) teaching level and (ii) learning phase [23]. The teacher aims to boost the median outcome of all students in the teaching process. The instructor will try to shift this means toward the best benefit for the individual.

Teacher Phase:

The instructor/teacher tries to improve the mean outcome of all students at this stage. Suppose the number of subjects (i.e. factors to be optimized) is "d" and 'N is the number of pupils i.e. population size $j = 1, 2, \ldots N$) and at k^{th} iteration, the mean student (μ^k) result, and teacher will aim to move this means to the best value (X_{best}^k). The difference is given by the average and the ideal average as:

$$Diff\_mean^k = rand\,(X_{best}^k - Tf\mu^k) \qquad (15.34)$$

where Tf is learning factors can be either 1 or 2 as: $Tf = round\,(1+\ rand\ (2\text{-}1))$.
Based on this difference the parameter values are updated as:

$$X\_teacher_i^k = X_{old,i}^k + Diff\_mean^k) \qquad (15.35)$$

Now, measure the value of the fitness function value based on the $X\_teacher_i^k$, and this value is compared with the value of fitness function of $x_{old,i}^k$ (i.e. previous value of particles) of the fitness function to produce a new population for the next step as:

$$x_{new,i}^k = \begin{cases} X\_teacher_i^k & \text{if } f(X\_teacher_i^k) \le f(x_{old,i}^k) \\ x_{old,i}^k & \text{if } f(x_{old,i}^k) < f\,(X\_teacher_i^k) \end{cases} \qquad (15.36)$$

This $x_{new,i}^k$ value plays the role of input for the learner phase of ith particle (student).
Learner Phase:
In this step, by engaging randomly with j^{th} students, i^{th} student increases his/her awareness so that: $x_{new,i}^k \ne x_{new,j}^k$. Now if fitness function value correspond to $x_{new,i}^k$ (i.e. fitness function value of ith student, $(f(x_{new,i}^k)$ is less than the fitness function value of jth student, $f(x_{new,j}^k)$ then particle values are updated as:

$$X\_learner_i^k = x_{new,i}^k + rand\,(x_{new,i}^k - x_{new,j}^k) \qquad (15.37)$$

Otherwise,

$$X\_learner_i^k = x_{new,i}^k + rand\,(x_{new,j}^k - x_{new,i}^k) \qquad (15.38)$$

The parameter values for the next iteration are obtained using following equation:

$$x_i^{k+1} = \begin{cases} X\_learner_i^k & \text{if } f(X\_learner_i^k) \le f(x_{new,i}^k) \\ x_{new,i}^k & \text{if } f(x_{new,i}^k) < f\,(X\_learner_i^k) \end{cases} \qquad (15.39)$$

15.4 OVERVIEW AND SOLUTION METHODOLOGY USING HYBRID STATISTICAL TRACKED PARTICLE SWARM OPTIMIZATION (STPSO)

15.4.1 Overview of Hybrid STPSO

The objectives of AGC can achieve through bio-Inspired evolutionary optimization techniques which have become a promising alternative to traditional mathematical techniques to search large solution space to attain the best solution to an engineering problem described by a nonlinear fitness function with a huge number of variables.

These algorithms are probabilistic pursuit methods that simulate biological characteristic development. The conduct of randomly created beginning seed is guided by learning, adoption, and development in extensive inquiry space to reach a close solution for the issue. To get better quality of the result, many bio-inspired algorithms have been proposed, which avoid the local trap and have faster convergence.

The optimization of a complex system is a troublesome undertaking. It can be effectively eliminated by joining capable optimization methods and investigation tools. A considerable lot of the known methods may not be acclimated to the arrangement of complex issues. This issue can be overwhelmed by new or comprehensive strategies which are even ready to manage complex issues inside the enormous search domains. A promising methodology for the development of these techniques is an intelligent combination of various advancement strategies by keeping up the helpful attributes of every strategy or change by some new parameters known as hybridization. In hybridization, the primary enhancement calculation might be work the entire time and the second method would just be changed now and again as indicated by a few conditions, or might be both the strategies would be connected with the other hand as per a fixed schedule or on the premise of some paradigm.

Various complex issues can be explained effectively utilizing such simple hybridization techniques. The intricate systems, as often as possible, require the utilization of in excess of two improvement techniques or some adjustment. For the solution of various applications, it is still extremely hard to choose which of the accessible evolutionary algorithms is suited and how the parameters and administrators of methods ought to be joined [24].

Many researchers have modified PSO algorithms in the last few years and all these variations have helped to converge quickly, but a multi-dimensional problem still takes a huge number of runs to obtain an optimum value. The current study uses certain statistical parameters as a tracker with less iteration to search for the world's best value.

Since basic PSO works properly with some of the easy problems of optimization, but for high dimensional multimodal optimization problems, it can get trapped on local best value. When convergence speed becomes too slow during the implementation of the basic PSO algorithm and then the suggested algorithm uses certain mathematical parameters to speed up particle velocity, and no further improvement in particle velocity is detected (like mean and standard deviation).

The proposed hybrid STPSO uses the following modified relations to update the velocity of particles:

$$v_{new,i}^{k+1} = wv_i^k + c_1 r_1 (\boldsymbol{pbest}_i^k - x_i^k) + c_2 r_2 (\boldsymbol{gbest}^k - x_i^k) + \boldsymbol{\beta}^k \quad (15.40)$$

Where β^k, a constant of acceleration that is centered on one of the mathematical parameters is represented. To compute this factor, the mean and the standard deviation of particle positions are used. The motive behind the use of these mathematical parameters is to boost the group behavior of particles since the mean value changes velocity in the direction of the best location by the reciprocal influence of various neighboring particles and standard deviation uses the influence of dispersion around the particle's average location. The following expressions are used to calculate the acceleration factor.

$$\beta^k = rand. \, (\boldsymbol{gbest}^k - X_s^k) \tag{15.41}$$

Where, $\boldsymbol{gbest}^k$ = position corresponding to the global best value.

X_s^k is computed using one of the statistical parameters:

1. Mean of particle position:

$$X_s^k = \bar{X} = \frac{\Sigma X_i}{N} \tag{15.42}$$

2. Standard deviation of particle position:

$$X_s^k = \sigma_s^k = \sqrt{\frac{\Sigma (X_i^k - \bar{X})^2}{N - 1}} \tag{15.43}$$

The position of the particle is updated using the following equation:

$$x_{new,i}^{k+1} = x_i^k + v_{new,i}^{k+1} \tag{15.44}$$

This chapter discusses hybrid STPSO using two mathematical criteria to look for an optimal value of increased variety. The first parameter is the mean value ($\bar{X}$) and the second is the default variance value (σ) of the position of the particle. If no change in the global best value of simple PSO is observed, then statistical tracking is added based on the mean value, and when mean tracking is slow, the particle position update is switched to standard deviation tracking.

Now, calculate fitness function value corresponding for $x_{new,i}^{k+1}$. Now particle is selected for next iteration as:

$$x_i^{k+1} = \begin{cases} x_{new,i}^{k+1} & \text{if } f(x_{new,i}^{k+1}) \le f(x_i^k) \\ x_i^k & \text{if } f(x_i^k) < f(x_{new,i}^k) \end{cases} \tag{15.45}$$

Figure 15.8 depicts modification in position in the proposed method:

15.4.2 COMPUTATIONAL ALGORITHM USING HYBRID STPSO METHOD

Step 1: Initially set the optimization parameters like population size (N), number of maximum iteration (itermax), number of variables (d), limits of search space (x_{min} and x_{max}), etc.

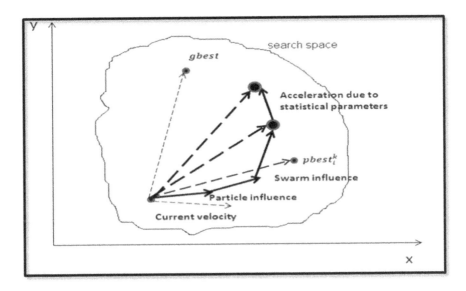

FIGURE 15.8   Vector representation of position updation in hybrid STPSO.

Step 2: Set k=1 further randomly generate particle's position and velocities within a specified search space using Equations 15.19 and 15.20, respectively.
Step 3: Evaluate the fitness function corresponding to each particle and then obtain $pbest_i^k$ and $gbest^k$.
Step 4: Update velocity and position of each particle using Equations 15.21 and 15.22, respectively.
Step 5: Evaluate a fitness function for each updated particle.
Step 6: Obtain $pbest_i^{k+1}$ and $gbest^{k+1}$.
Step 7: If $|f(gbest^k) - f(gbest^{k+1})| \leq \varepsilon$ then go to step-9. Where ε is a switch criterion (set at 0.0001) to change tracks from basic PSO to mean.
Step 8: Increase the generation count $k=k+1$, and go to step-4.
Step 9: Calculate the mean of particle position $\bar{X}$ and determine the acceleration factor using Equation 15.41.
Step 10: Update velocity and position of each particle using Equations 15.44 and 15.45, respectively.
Step 11: Calculate fitness function further for updated particles and obtain $pbest_i^k$ and $gbest^k$.
Step 12: If $|f(gbest^k) - f(gbest^{k+1})| \leq \varepsilon$ then go to step-14.
Step 13: Increment the generation count $k=k+1$ and go to step-9.
Step 14: Calculate standard deviation vector σ using Equation 15.43.
Step 15: Update each particle using Equations 15.44 and 15.45.
Step 16: Set $k=k+1$. If $k \geq itermax$ then terminate the process otherwise repeat from step-11.

The flow chart representation of the proposed hybrid STPSO is shown in Figure 15.9

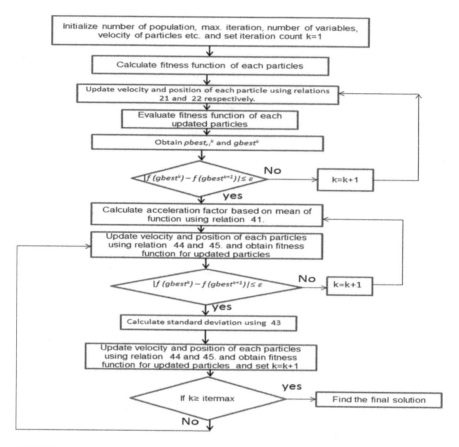

FIGURE 15.9 Flow chart representation of hybrid STPSO.

Performance of the proposed hybrid STPSO is evaluated by implementing it on a different test system as follows:

15.5 IMPLEMENTATION ON BENCHMARK FUNCTIONS

Firstly, as set out in Table 15.1, the performance of the proposed algorithm is checked on eight benchmark test functions [137]. To illustrate the benefit of the suggested modification in PSO, this chapter takes multiple dimensions (D_d), for example, 5, 10, and 20. The multimodal function has several local minimum values, such as the Griewank function. The noisy function has random noise, so the objective function in consecutive iterations is altered. The Rosenbrock function has a broad search space with the lowest global value, so it is impossible to search for the best value. The De-Jong function is a basic two-dimensional reference element. For different dimensions, the first three benchmark functions have the same parameter values. (i.e. $F_{min} = 0$; for $x = 0, 0, ...$). Shifted similar functions (functions 5 and 6) transfer the optimum point to a random location to prevent this problem, and establish different global optimal parameter values for different dimensions. Hybrid

TABLE 15.1

Description of Benchmark Function [25]

Function Name	Function Design	Range (Search Space)	Global Value
Griewank	$fitness\ fun1 = \frac{1}{4000}\sum_{i=1}^{D}x_i^2 - \prod_{i=1}^{D}\cos\left(\frac{x_i}{\sqrt{i}}\right) + 1$	[−50, 50]	$F_{min} = 0$; x = (0, 0 …)
Noisy	$fitness\ fun2 = \sum_{i=0}^{D}(i+1)x_i^4 + rand(0,1)$	[−1.28, 1.28]	$F_{min} = 0$;
Rosenbrock	$fitness\ fun3 = \sum_{i=0}^{D}100(x_{i+1} - x_i^2)^2 + (1-x_i)^2$	[−30, 30]	$F_{min} = 0$; x = (1, 1, …)
De-Jong	$max(fitnessfun4) = 3905.93 - 100(x_1^2 - x_2)^2 - (1-x_1)^2$	[−2.048, 2.048]	$F_{min} = 3905.93$; x = (1, 1)
Shifted sphere	$fitness\ fun5 = \sum_{i=1}^{D}(z_i^2 + f_{-bias1})$, z = x − o; where o is randomly shifted global point and $f_{-bias}1 = -450$	[−100, 100]	$F_{min} = f\_bias1X = 0$
Shifted rotated rastrigin	$fitness\ fun6 = \sum_{i=1}^{D}z_i^2 - 10\cos(2\pi z_i) + 10 + f\_bias2$, z = (x − o) * M; where o is randomly shifted global point, M: linear transformation matrix and $f_{-bias}2 = -330$	[−5.12, 5.12]	$F_{min} = f\_bias2X = 0$
Hybrid composition	$fit\ func7, F(x) = \sum_{i=1}^{n'}(w_i * ((\frac{f_i(x-o_i)}{\lambda_i} * M_i + bias_i)) + f_{bias}3, where, n' = 10;$ f_bias3 = $120f_{1-3}(x) = $ Rastrigin function for $D_d = 10, f_{4-6}(x) = $ Griewank function for $D_d = 10, f_{7-8}(x) = $ Ackley function for $D_d = 10, f_{9-10}(x) = $ Sphere function for $D_d = 10, \sigma_i = 1$ for i = 1, 2 … 10 $\lambda_i = [1,1,10,10,5/60,5/32,5/32,5/100,5/100]$	$f_{1-3} = [−5.12, 5.12], f_{4-6} = [−50, 50] f_{7-8} = [−32,32] f_{9-10} = [−100,100]$	$F_{min} = f\_bias3X = 0$
Hybrid composition with noise	$fit\ func8, F(x) = fit\ func10 * (1 + 0.2.* (rand(0,1)))$ All settings are the same as hybrid composition function	Same as hybrid composition function	$F_{min} = f\_bias3X = 0$

composition test functions (functions 7 and 8 in Table 15.1) are used to maximize system complexity, in which randomly located global optima and several randomly located local optima are considered [25].

15.5.1 PERFORMANCE EVALUATION

The performance evaluation of hybrid STPSO is tested on standard benchmark functions as tabulated in Table 15.1 and compared it with other variables of PSO and PSO with mean and standard deviation. The specified number of the maximum iteration for each algorithm is 300 (for the reference function of sizes D = 5 and 10), 400 for the reference function of sizes D = 20, and 300 for the hybrid method.

Figure 15.10 visualizes the convergence of the Griwank function in successive iterations using hybrid STPSO in MATLAB. It clearly shows that particles in the first iteration are randomly distributed in search space and as the number of iterations increases, they converged at one point that is the global point only in the 18th iteration using hybrid STPSO. The relation of the convergence for the best fitness value with respect to the number of iterations for the test functions is seen in Figure 15.11. These figures demonstrate the hybrid STPSO, seeks the right

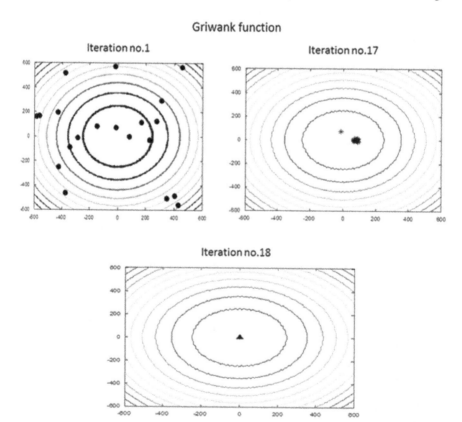

FIGURE 15.10 Convergence of Griwank function.

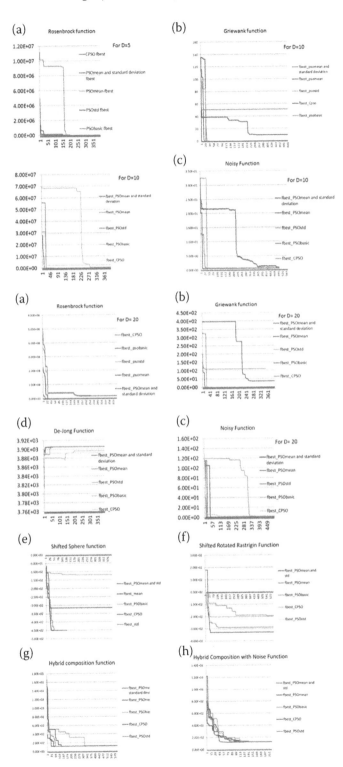

FIGURE 15.11 Convergence for best fitness value with respect to iteration count for different dimensions of Rosenbrock (A), Griwank function (B), Noisy function (C), De-Jong (D), Shifted sphere (E), Shifted Rotated Rastrigin (F), Hybrid composition (G), and Hybrid composition with Noise (H).

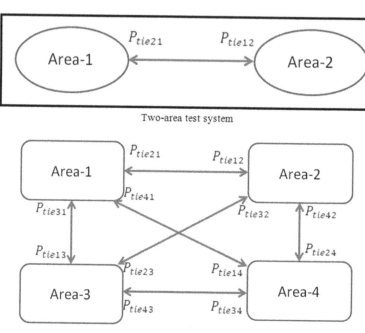

Two-area test system

Four-area test system

FIGURE 15.12 Schematic diagram of two-area and four-area test system.

values faster than other approaches for dynamic multimodal, and shifted rotated test functions with mean and standard deviations.

15.6 IMPLEMENTATION ON PROPOSED AUTOMATIC GENERATION CONTROL SYSTEMS

In the proposed research, hybrid STPSO is implemented for two test systems as two-area and four-area linear and nonlinear AGC systems. A schematic diagram of the proposed test system is shown in Figure 15.12.

The objectives of AGC problem are to obtain parameter so that desired system performance like minimum overshoot, undershoot, and settling time are obtained. The present research has used following optimization function to achieve the above goals [50].

$$\textit{fitness function} = \int_0^t \sum_{j=1}^n (\Delta f_j)^2 + \sum_{j=1}^{n-1} (\Delta P_{tie,j})^2 dt \qquad (15.46)$$

This fitness function is minimized subject to the following constraints in state space equation

$$\sum_{i=1}^{n} \Delta P_{tie,i} = 0 \tag{15.47}$$

$$\Delta P_{tie,n} = -\sum_{i=1}^{n-1} \Delta P_{tie,i} \tag{15.48}$$

Substituting this condition and modifying state matrix of Equations (15.12) as follows: $x = [\Delta f_1, \Delta P_{E,1}, \Delta P_{T,1}, \Delta \dot{P}_{T,1}, \Delta P_{c,1}, \Delta f_2, \Delta P_{E,2}, \Delta P_{T,2}, \Delta \dot{P}_{T,2}, \Delta P_{c,2}, \Delta P_{tie,1}]^T$ for two-area system and

$$x = \begin{bmatrix} \Delta f_1, \Delta P_{E,1}, \Delta P_{T,1}, \Delta \dot{P}_{T,1}, \Delta P_{c,1}, \Delta f_2, \Delta P_{E,2}, \Delta P_{T,2}, \Delta \dot{P}_{T,2}, \Delta P_{c,2}, \Delta f_3, \Delta P_{E,3}, \Delta P_{T,3}, \\ \Delta \dot{P}_{T,3}, \Delta P_{c,3}, \Delta f_4, \Delta P_{E,4}, \Delta P_{T,4}, \Delta \dot{P}_{T,4}, \Delta P_{c,4}, \Delta P_{tie,1}, \Delta P_{tie,2}, \Delta P_{tie,3}, \Delta P_{tie,4}] \end{bmatrix}^T$$

In which $\Delta P_{tie,1} = 2\pi T_{12}(\Delta f_1 - \Delta f_2)$ is positive for area-1 and negative for area-2 in two-area system. Similarly, $\Delta P_{tie,1}, \Delta P_{tie,2}, \Delta P_{tie,3}$ and $\Delta P_{tie,4}$ for four-areas system can be written as:

$$\Delta P_{tie,1} = (2\pi T_{12} + 2\pi T_{13} + 2\pi T_{14})\Delta f_1 - 2\pi T_{12}\Delta f_2 - 2\pi T_{13}\Delta f_3 - 2\pi T_{14}\Delta f_4)$$
$$\Delta P_{tie,2} = (2\pi T_{12} + 2\pi T_{23} + 2\pi T_{24})\Delta f_2 - 2\pi T_{12}\Delta f_1 - 2\pi T_{23}\Delta f_3 - 2\pi T_{24}\Delta f_4)$$
$$\Delta P_{tie,3} = (2\pi T_{32} + 2\pi T_{31} + 2\pi T_{34})\Delta f_3 - 2\pi T_{32}\Delta f_2 - 2\pi T_{31}\Delta f_1 - 2\pi T_{34}\Delta f_4)$$

And

$$\Delta P_{tie,4} = -\Delta P_{tie,1} - \Delta P_{tie,2} - \Delta P_{tie,2} \tag{15.49}$$

The state transition matrix and [U] matrix for two-area system become:

$$[A] = \begin{bmatrix}
-\frac{1}{T_{p1}} & 0 & \frac{K_{p1}}{T_{p1}} & 0 & 0 & 0 & 0 & 0 & 0 & 0 & -\frac{K_{p1}}{T_{p1}} \\
a_{21} & -\frac{1}{T_{g1}} & 0 & 0 & \frac{1}{T_{g1}} & 0 & 0 & 0 & 0 & 0 & 0 \\
0 & 0 & 0 & 1 & 0 & 0 & 0 & 0 & 0 & 0 & 0 \\
a_{41} & \left(\frac{1}{T_{r1}T_{r1}} - \frac{K_{r1}}{T_{g1}T_{r1}}\right) & \frac{-1}{T_{r1}T_{r1}} & -\left(\frac{T_{r1}+T_{r1}}{T_{r1}T_{r1}}\right) & \frac{K_{r,1}}{T_{g1}T_{r1}} & 0 & 0 & 0 & 0 & 0 & 0 \\
-K_{I1}B_1 & 0 & 0 & 0 & 0 & 0 & 0 & 0 & 0 & 0 & -K_{I1} \\
0 & 0 & 0 & 0 & 0 & -\frac{1}{T_{p2}} & 0 & \frac{K_{p2}}{T_{p2}} & 0 & 0 & -\frac{K_{p2}}{T_{p2}} \\
0 & 0 & 0 & 0 & 0 & a_{76} & -\frac{1}{T_{g2}} & 0 & 0 & \frac{1}{T_{g2}} & 0 \\
0 & 0 & 0 & 0 & 0 & 0 & 0 & 0 & 1 & 0 & 0 \\
0 & 0 & 0 & 0 & 0 & a_{96} & \left(\frac{1}{T_{I2}T_{r2}} - \frac{K_{r2}}{T_{g2}T_{I2}}\right) & \frac{-1}{T_{r2}T_{I2}} & -\left(\frac{T_{r2}+T_{I2}}{T_{r2}T_{I2}}\right) & \frac{K_{r,2}}{T_{g2}T_{I2}} & 0 \\
0 & 0 & 0 & 0 & 0 & -K_{I2}B_2 & 0 & 0 & 0 & 0 & -K_{I2} \\
2\pi T_{12} & 0 & 0 & 0 & 0 & -2\pi T_{12} & 0 & 0 & 0 & 0 & 0
\end{bmatrix}$$

$$\tag{15.50}$$

$$[U] = \left[\frac{-K_{p1}}{T_{p1}} \Delta P_{D1}, \ U_2, \ 0, \ U_4, \ 0, \ \frac{-K_{p2}}{T_{p2}} \Delta P_{D2}, \ U_7, \ 0, \ U_9, \ 0, \ 0 \right]^T \quad (15.51)$$

Where, a_{21}, a_{41}, a_{76}, a_{96}, U_2, U_4, U_7 and U_9 find out using Equations 15.15–15.18. Similarly, the above matrix can be written for four-area system.

15.6.1 Performance Evaluation

In this section, the performance of a hybrid STPSO has been compared with basic PSO, CPSO, DE, BB–BC, and TLBO algorithms. For comparison purposes, 100 runs have been executed for each technique.

Figure 15.13 depicts that hybrid STPSO also gives minimum fitness function value compared to the PSO, CPSO, DE, BB–BC, and TLBO for two-area AGC systems with both linear and nonlinear systems.

This chapter evaluates the optimal controller for AGC of two-area and four-area systems with and without considering nonlinearity in the system. All the constants, time constants, and transfer function gains are

$$k_{g,1} = k_{g,2} = k_{g,3} = k_{g,4} = 1, T_{g,1} = T_{g,2} = T_{g,3} = T_{g,4} = 0.08,$$
$$k_{t,1} = k_{t,2} = k_{t,3} = k_{t,4} = 1, T_{t,1} = T_{t,2} = T_{t,3} = T_{t,4} = 0.3,$$
$$K_{r,1} = K_{r,2} = K_{r,3} = K_{r,4} = 0.5, T_{r,1} = T_{r,2} = T_{r,3} = T_{r,4} = 10,$$
$$K_{p,1} = K_{p,2} = K_{p,3} = K_{p,4} = 120, T_{p,1} = T_{p,2} = T_{p,3} = T_{p,4} = 20,$$
$$\beta_1 = \beta_2 = \beta_3 = \beta_4 = 0.872, R_1 = R_2 = R_3 = R_4 = 2.2568,$$
$$2\pi T_1 = 2\pi T_2 = 2\pi T_3 = 2\pi T_4 = 0.05,$$

The present chapter illustrates different heuristics algorithms for these two test systems. The effectiveness of the proposed hybrid STPSO is illustrated by comparing it with other evolutionary algorithms like PSO, CPSO, DE, BB–BC, and TLBO for these systems.

In this chapter, an integral controller is considered for both test systems. The optimal value of controller gain and frequency bias constant is evaluated by evolutionary algorithms by considering identical values for analysis purposes that is, $K_{I1} = K_{I2} = K_{I3} = K_{I4}$ and $B_1 = B_2 = B_3 = B_4$. The optimum values for two test systems corresponding to performance indices considered are summarized in Table 15.2(A) and Table 15.2(B).

Figure 15.14 illustrated that change in frequency of both areas (area-1, area-2) have better performance parameters in terms of overshoot, steady state error, and settling time for hybrid STPSO compared with other heuristic algorithms without considering the dead band.

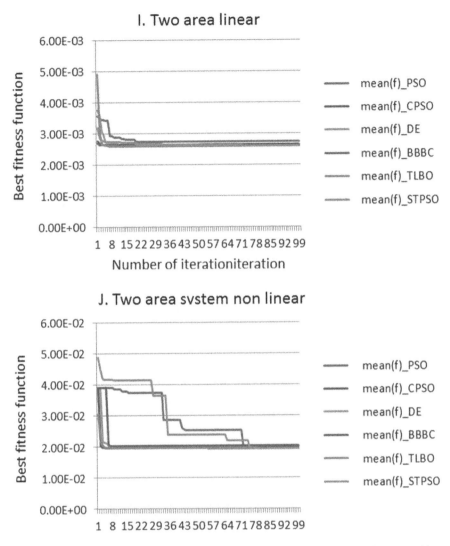

FIGURE 15.13 Convergence profile for two-area (I-linear, nonlinear) AGC system of best mean fitness function value with respect to iteration count.

This chapter applies proposed hybrid STPSO algorithms to investigate the effect of a dead band on dynamic performance. The magnitude of the dead band is taking 0.06% in each area. Since the change in frequency varies slowly with time, so can assume that during any particular time interval each area operates entirely inside the dead band or outside it. Therefore, the system can be considered a piecewise linear system with consideration of the governor dead band. With the inclusion of the effect of a dead band, still, the hybrid STPSO gives a satisfactory dynamic response to a change in frequency of all areas compared with another algorithm as shown in Figure 15.15. In this figure, the dotted

TABLE 15.2A

Parameters Value for Test System Without Considering Dead Band

Test System	Algorithms	Parameters		Fitness Function
		K_I	B	
Two-area AGC	PSO	0.15159586	2.0386451	0.0025936321
	CPSO	0.1471251	2.0839265	0.0025931256
	DE	0.10929973	2.8573298	0.0025889345
	BB-BC	0.30990689	1.0278501	0.0026245685
	TLBO	0.069759478	4.4881282	0.0025875135
	Hybrid STPSO	0.078418113	3.09937178	0.0025874981
Four-area AGC	PSO	0.32203068	0.9660046	0.0054784203
	CPSO	0.3515	0.8800	0.0054350498
	DE	0.3451	0.9000	0.0054300418
	BB-BC	0.4239	0.7342	0.0054867217
	TLBO	0.3580	0.8977	0.0054330442
	Hybrid STPSO	0.3467	0.8993	0.0053300418

TABLE 15.2B

Parameters Value for Test System with Dead Band (Nonlinear System)

Test System	Algorithms	Parameters		Fitness Function
		K_I	B	
Two-area AGC	PSO	0.33858111	0.99996474	0.01947673
	CPSO	0.3133572	0.87027290	0.019566644
	DE	0.30757	0.9000	0.019425808
	BB-BC	0.43858	0.8918	0.020240716
	TLBO	0.2260	0.8913	0.019541211
	Hybrid STPSO	0.37747968	0.89128049	0.019433454
Four-area AGC	PSO	0.2746904	0.87208495	0.0088929749
	CPSO	0.2656	0.8086	0.0087297675
	DE	0.2386	0.8949	0.008708383
	BB-BC	0.29568052	0.88709475	0.0088690639
	TLBO	0.23619314	0.89047862	0.0087363357
	Hybrid STPSO	0.23480	0.8106	0.0086047532

(a)

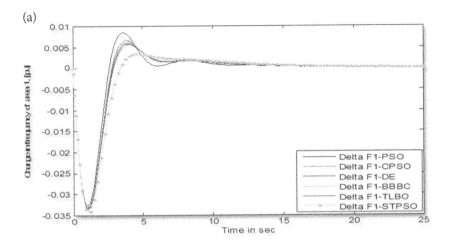

(b)

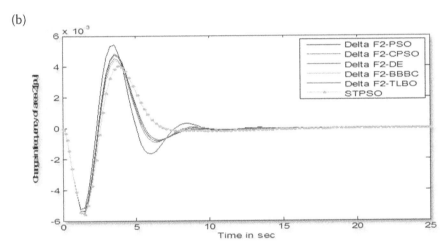

FIGURE 15.14 Comparison of hybrid STPSO with other evolutionary algorithms for change in frequency of area-1(a) and area-2 (b) with respect to time.

yellow curve shows that change in frequency of area-1 (Figure 15.15(e)) and area-2 (Figure 2.16(f)) have better performance parameters in terms of the maximum deviation from the desired value and selling time using the hybrid STPSO method.

With the increase in size and a number of variables, still, the proposed hybrid STPSO gives a better performance compared to various other algorithms. Figure 15.16 illustrates the response of change in frequency of area-1 (Figure 15.16(a)), area-2 (Figure 15.16(b)), area-3 (Figure 15.16(c)), and area-4 (Figure 15.6(d)), respectively. These figures clearly analyzed that the response obtained by hybrid STPSO, which is shown by the dotted blue line, has a relatively better performance compared to the response obtained by the PSO, CPSO, DE, BB–BC, and TLBO methods.

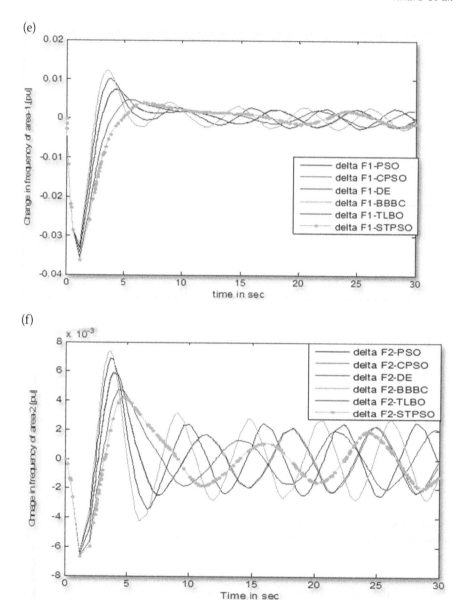

FIGURE 15.15 Comparison of hybrid STPSO with other evolutionary algorithms for change in frequency of area-1(e) and area-2 (f) with respect to time.

The inclusion of dead band's complexity in the system is increased in terms of a large number of parameters to be optimized due to a number of areas as well as nonlinearity. Further, hybrid STPSO performs better and gives a satisfactory dynamic response. Figure 15.17 shows a comparative analysis of the dynamic performance of area-1 (Figure 15.17(I)), area-2 (Figure 15.17(J)), area-3 (Figure 15.17(k)), and area-4 (Figure 15.17(L)) for the best controller gains

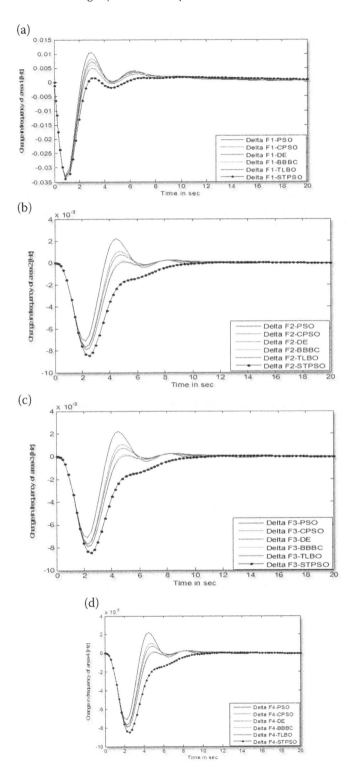

FIGURE 15.16 Comparison of change in frequency of area-1(a), area-2(b), area-3(c), area-4(d) with respect to time.

among 100 runs of each algorithm. These figures clearly illustrate the response from hybrid STPSO-based optimized AGC with the dead band (shown by dotted yellow curve) still gives good performance characteristics compared to another algorithm.

Comparison Based on Statistical Evaluation

Table 15.3 gives the comparative result for AGC for two-area and four-area systems. This table illustrates that the hybrid STPSO gives improved results compared with other evolutionary algorithms (PSO, CPSO, DE, BB–BC, and TLBO) for both linear and nonlinear frequency control systems. This table shows that statistical performance of hybrid STPSO has better statistical parameters compared with other heuristic algorithms in terms of minimum fitness function (two-area linear: PSO-0.00263, CPSO-0.00262, DE-0.00262, BB–BC-0.00261, TLBO-0.00260, hybrid STPSO-0.00259, two-area nonlinear: PSO-0.0195, CPSO-0.0196, DE-0.0195, BB–BC-0.0202, TLBO- 0.0195, STPSO-0.0194, four-area linear: PSO-0.00547, CPSO-0.00545, DE-0.00545, BB–BC-0.00546, TLBO: 0.00543, STPSO-0.00534, four-area nonlinear: PSO-0.00871, CPSO-0.00871, DE-0.00871, BB–BC-0.00881, TLBO-0.00870, STPSO-0.00861), mean of the fitness function, maximum of the fitness function, etc. This table also evaluates that for less number of optimal parameters (in two-area) and linear system the statistical parameters are approximately the same but as the number of optimal parameters increases, linear and nonlinear STPSO still give better statistical parameter value compared with other heuristic algorithms.

15.7 CONCLUSION

An interconnected power system has various areas of generating units, which are associated through tie-lines. The effective task of an interconnected power system requires the coordinating of an aggregate generation with load demand and related system losses at an ostensible frequency and terminal voltage. A nominal frequency is achieved by maintaining the generation and demand of real power. On a large scale, practical system real power requirement is continuously changing, therefore manual regulation is not feasible. Another issue is that the interconnection of the power systems brings about tremendous increments in both the order of the system and the quantity of the tuning controller parameters. For this reason, AGC (or load frequency control (LFC)) is outlined and executed to naturally adjust generated power and load demand in every area. The AGC responds to a frequency signal via the speed governor and control valves. One of the advantages of AGC is providing real-time control to match the change in a generation to the change in load while maintaining tie-line flows within a tolerance band and good tracking load demand with disturbances.

A number of control strategies e.g. integral, proportional integral (PI), and proportional integral derivative (PID) have been developed. Due to a fixed structure and constant parameter, the conventional controller maintains system frequency for

(i)

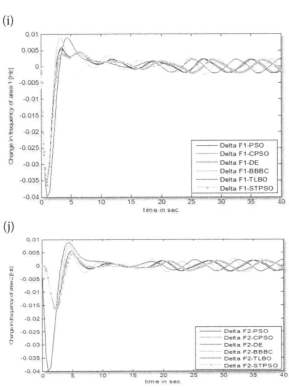

(j)

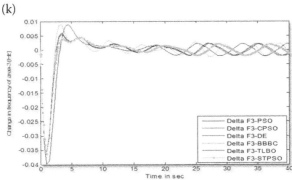

(k)

(l)

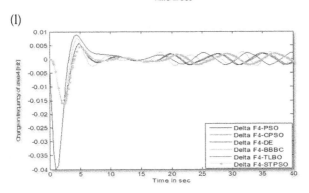

FIGURE 15.17 Comparison of hybrid STPSO with other evolutionary algorithms for change in frequency of area-1(I), area-2 (J), area-3(K), area-4(L).

one operating condition. Since most of the power system components are nonlinear, this controller may not able to provide desired performance for other operating conditions. To eliminate the above problem, various decentralized load frequency controller was produced yet they require data on system states. Aside from evaluations in control concepts, there have been numerous progressions amid the most recent decade, for example, deregulation of the power industry and utilization of superconducting magnetic energy storage, wind turbines, photovoltaic cells as the non-conventional energy sources of electrical energy to the system. Therefore, the control strategy associated with an AGC problem has changed.

An updated PSO algorithm has been built in this chapter by adding statistical parameters to find the best global value. The proposed hybrid STPSO strategy is contrasted with existing optimization methods, such as simple PSO, CPSO, PSO with mean tracking, and PSO with standard deviation tracking in the standard test function. The additional efficiency of the STPSO hybrid is compared to the PSO, CPSO, DE, BB–BC, and TLBO AGC for the linear and nonlinear systems. Experimental findings show that hybrid STPSO performs best in terms of the convergence profile and statistical parameters for different complex shift rotations with a noise effect. Numerical findings indicate that the suggested approach will efficiently avoid the trapping of the best local benefit.

15.7.1 Guide Line for Future Work

The usage of the solar wind sustainable power source system is expanding step by step and has indicated enormous development in the most recent couple of decades for power generation everywhere throughout the world. With the improvement of new advances in the field of frequency control of traditional AGC and solar wind hybrid sustainable power system, another issue emerges, which turns out to be significantly more fascinating to be settled. These issues will be remunerated by some future research in the particular field.

- Some issue is accounted for to discover the exact area and atmosphere condition, site to site information is required, which is hard to get to a remote area. Consequently, it is important to build up optimization methods and geological programming to discover the capability of solar radiation and wind speed to keep up system frequency.
- The concept of game theory will also apply to AGC to find out the strategic decision-making process.
- Game theory is a current theme in the field of a non-conventional energy system. It is important to utilize different monopoly and duopoly ideas to discover the correct system cost and rescue estimation of the incorporated system in the perspective of the expanded step-by-step establishment of the sustainable power source system.

TABLE 15.3

Statistical Comparison of Two and Four-Area (linear and nonlinear) AGC with Confidence Level 0.95

System	Algorithm	Mean (fbest) (μ)	Median (fbest)	std (fbest) (σ)	Best Value (fbest)	Max. Value (fbest)	Freq. of Convergence	Standard Error of the Mean Fitness Function (ε) = $C*\sigma/\sqrt{n}$; n = pop size; C= 97.5^{th} Percentile of Distribution	Length of Confidence $2*C*\varepsilon$
Two-area (linear)	PSO	0.0028	0.0026	0.00808	0.0026	0.0029	69	$14.383*10^{-4}$	0.002.8047
	CPSO	0.0026	0.0261	0.00001	0.0026	0.0026	65	$1.8*10^{-6}$	0.0000035
	DE	0.0026	0.0026	0.00001	0.0026	0.0026	80	$1.8*10^{-6}$	0.0000035
	BB-BC	0.0026	0.0026	0.00001	0.0026	0.0026	74	$2.1*10^{-6}$	0.0000042
	TLBO	0.0026	0.0026	0.00001	0.0026	0.0027	83	$2.1*10^{-6}$	0.0000042
	PSOmean and std (STPSO)	0.0026	0.0026	0.00004	0.0026	0.0026	92	$7.3*10^{-6}$	0.0000142
Two-area (non linear)	PSO	0.0195	0.0195	0.000022	0.0195	0.0195	82	$3.9*10^{-6}$	0.0000076
	CPSO	0.0196	0.0196	0.000002	0.0196	0.0196	75	$0.4*10^{-6}$	0.0000007
	DE	0.0194	0.0194	0.000013	0.0195	0.0194	83	$1.8*10^{-6}$	0.0000035
	BB-BC	0.0202	0.0202	0.000182	0.0202	0.0202	40	$3.24*10^{-4}$	0.0000632
	TLBO	0.0195	0.0195	0.000038	0.0195	0.0196	84	$6.8*10^{-6}$	0.0000132
	PSOmean and std (STPSO)	0.01943	0.0194	0.000012	0.0194	0.0195	85	$2.1*10^{-6}$	0.0000042

(Continued)

TABLE 15.3 (Continued)

Statistical Comparison of Two and Four-Area (linear and nonlinear) AGC with Confidence Level 0.95

System	Algorithm	Mean (fbest) (μ)	Median (fbest)	std (fbest) (σ)	Best Value (fbest)	Max. Value (fbest)	Freq. of Convergence	Standard Error of the Mean Fitness Function (ε) = $C*\sigma/\sqrt{n}$; n = pop size; C= 97.5^{th} Percentile of Distribution	Length of Confidence $2*C*\varepsilon$
Four-area (linear)	PSO	0.00548	0.00542	0.00014	0.00547	0.00585	78	0.0000247	0.0000481
	CPSO	0.00544	0.00544	0.00002	0.00545	0.00548	50	0.0000032	0.0000063
	DE	0.00543	0.00543	0.00000	0.00545	0.00543	77	0.0000000	0.0000000
	BB-BC	0.00549	0.00549	0.00001	0.00546	0.00551	65	0.0000018	0.0000036
	TLBO	0.00543	0.00543	0.00001	0.00543	0.00543	86	0.0000001	0.0000001
	PSOmean and std (STPSO)	0.00533	0.00535	0.00007	0.00534	0.00537	88	0.0000131	0.0000255
Four-area (non linear)	PSO	0.00889	0.00893	0.00014	0.00871	0.00913	48	0.00003	0.00005
	CPSO	0.00873	0.00872	0.00003	0.00871	0.00886	62	0.00001	0.00001
	DE	0.00871	0.00871	0.00013	0.00871	0.00871	81	0.00002	0.00005
	BB-BC	0.00887	0.00885	0.00005	0.00881	0.00894	42	0.00001	0.00002
	TLBO	0.00874	0.00874	0.00000	0.00870	0.00874	64	0.00000	0.00000
	PSOmean and std (STPSO)	0.00860	0.00878	0.00010	0.00861	0.00868	83	0.00002	0.00003

- It is important to expand centralized and multilevel controlling strategy which avoids the potential multifaceted nature of correspondence system and vast calculation load which is subjected to a single-point failure.

REFERENCES

1. O.I. Elgerd and C. Fosha, "Optimum Megawatt-frequency control of multiarea electric energy systems", *IEEE Transaction on Power Apparatus and Systems*, vol. PAS-89, no. 4, pp. 556–563, April 1970.
2. J. Nanda and B.L. Kaul, "Automatic generation control of an interconnected power system", *IEE proc vol. 125*, no. 5, pp. 385–391, May 1978.
3. O.I. Elgerd, *"Electric energy system theory: An introduction"*, Mcgraw Hill, United States, 1982.
4. H. Saadat, *"Power system analysis"*, McGraw-Hill, New Delhi, 2002.
5. S. Shubha, "Load frequency control with fuzzy logic controller considering governor dead band and generation rate constraint non-linearity", *World Applied Sciences Journal*, vol. 29, no. 8, pp. 1059–1066, 2014.
6. P. Kundur, *"Power system stability and control"*, McGraw- Hill, New Delhi, 2009.
7. G. Panda, S. Panda and C. Ardil, "Automatic generation control multi area electric energy system using modified GA", *World Academy of Science, Engg. and Technology*, vol. 39, pp. 717–725, 2010.
8. G. Panda, S. Panda and C. Ardil, "Automatic generation control of interconnected power system with generation rate constraints by hybrid neuro fuzzy approach", *International Journal of Electrical and Electronics Engg.*, vol. 2, NO. 1, pp. 3–9, 2009.
9. D.K. Chaturvedi, "Soft computing: Techniques and its application in electrical engineering" *Springer*, vol. 103, 2008.
10. S. Kumar Das, A. Kumar, B. Das and A.P. Burnwal, "On soft computing techniques in various area", *CS and IT- CSCP*, pp 59–68. 2013.
11. J. Kennedy and R. Eberhart, "Particle swarm optimization," *IEEE International Conf. Neural Networks*, vol. 4, Perth, Australia, pp. 1942–1948, 1995.
12. P. Acharjee, S.K. Goswami, "Chaotic particle swarm optimization based robust load flow", *Journal of Electrical Power and Energy System* vol. 32, pp. 141–148, 2010.
13. Y. Shi, "Particle swarm optimization: developments, applications and resources", In Proceedings of the 2001 congress on evolutionary computation (IEEE Cat. No. 01TH8546) (Vol. 1, pp. 81–86). IEEE, 2001, May.
14. L. dos Santos Coelho, and V. Cocco Mariani, "A novel chaotic particle swarm optimization approach using Hénon map and implicit filtering local search for economic load dispatch", *Journal, Chaos Solutions and Fractals*, vol. 39, pp. 510–518, 2009.
15. J. Zhang and X. Ding, "A multi swarm self adaptive and co-operative particle swarm optimization", *Engineering Application of Artificial Intelligence*, vol. 24, pp. 958–967, 2011.
16. R. Storn and K. Price," Differential evolution – A simple and efficient adaptive scheme for global optimization over continuous spaces", *Technical report TR-95-012*, 1995
17. S. Das and N. Suganhan Ponnuthurai, "Differential evolution: A survey of the state of the art", *IEEE Transaction on Evolutionary Computation*, vol. 15, No. 1, pp. 4–29, 2011.
18. K.E. Osman, and E. Ibrahim, "New optimization method: Big Bang-Big Crunch", *Advances in Engineering Software*, vol. 37, pp. 106–111, 2006.

19. A.O. Salau, Y.W. Gebru, and D. Bitew, "Optimal network reconfiguration for power loss minimization and voltage profile enhancement in distribution systems", *Heliyon*, vol. 6, No. 6, e04233, 2020. DOI: 10.1016/j.heliyon.2020.e04233

20. E. Yesil, and L. Urbas, "Big Bang- Big Crunch learning method for fuzzy cognitive maps", *World Academy of Science, Engg and Technology*, vol.4, no. 11, pp. 1756–1765, 2010.

21. P. Prudhvi, "A complete copper optimization technique using BB-BC in a smart home for a smarter grid and comparison with GA", *IEEE Conference (CCECE)*, pp. 69–72, 2011.

22. R.V. Rao, V.J. Savsaniand and D.P. Vakharia, "Teaching–learning-based optimization: An optimization method for continuous non-linear large scale problems", *Information Sciences*, vol. 183, pp. 1–15, 2012.

23. R. Venkata Rao and V. Patel, "An elitist teaching-learning-based optimization algorithm for solving complex constrained optimization problems", *International Journal of Industrial Engineering Computations,* vol. 3 pp. 535–560, 2012.

24. J. Hewlett and G. Dundar, "Merge of evolutionary computation with gradient based method for optimization problems", *IEEE Conference*, pp. 3304–3309, 2007.

25. P.N. Suganthan, N. Hasen, J.J. Liang, K. Deb, Y.P. Chen, A. Auger, and S. Tiwari, "Problem definition and evaluation criteria for the CEC 2005 special session on real parameter optimization", *Technical report, Nanyang Technology University*, Singapore, KanGAL report number 2005005, 2005.

16 Steganography and Steganalysis Using Machine Learning

Sachin Dhawan[1], Dr. Rashmi Gupta[2], and Arun Kumar Rana[3]

[1]Ambedkar Institute of Advanced Communication Technologies and Research, Geeta Colony, New Delhi
[2]NSUT, East Campus, New Delhi
[3]Panipat Institute of Engineering and Technology, Samalkha

16.1 INTRODUCTION

In the era of communication, the number of individuals using cell phones for communicating, working, and networking has increased considerably. It has become a necessity in our daily life and work. Yet, there are innumerable security issues associated with this advanced technology. For instance, personal information can be accessed, spread, used unlawfully, or copyrighted, and more. In addressing such issues, data stowing is being given a lot of consideration to enhance security and offer copyright. Data concealing implies that mystery data is covered up in the cover picture by using a few attributes of the cover picture, and during the time of transfer of the cover picture, no inconsistencies are found by the indicator so that the stego picture can be securely sent to the collector. The collector removes mystery data through a specific calculation to acknowledge mystery correspondence. Among these, mystery data can be a bit of text, a picture, etc. During the time spent stowing away, the typical strategy is to change over mystery data into a spot stream and conceal the touch data in the cover media. Moreover, the picture can be straightforwardly covered up in another picture, or the mystery data can be related to the planning word reference, and the data containing the planning articles can be sent to the collector. Steganography is a method for undercover correspondence and is considerably in demand in public security and military affairs [1,2]. Notwithstanding, steganography is likewise utilized by individuals for enhancing security of organization communication. Indeed, data covering up has likewise been utilized in surveillance, psychological oppressor assaults, violations, and different exercises in recent years. Under such conditions, how to adequately regulate steganography and forestall and block its pernicious or illicit application has become a critical need of military and security offices in different nations. Therefore, steganalysis has

DOI: 10.1201/9781003140443-16

been generally considered and created due to increase in data stowing. Steganalysis alludes to the cycle wherein the locator decides if the stego picture contains mystery data or not during the distribution of the stego image [3–5]. Steganography and steganalysis are two types of calculations that confine one another and restrict one another.

Machine learning is an area of AI that holds capacity to learn without being customized, whereas Deep Learning is a subset of AI [6–8]. In this part, steganography, steganalysis, and AI procedures are clarified and the cycle and opportunities for steganalysis in different AI systems are depicted. Some datasets on stego-pictures are arranged and the models for simulation are prepared [9–18].

16.2 STEGANOGRAPHY

Figure 16.1 shows that steganography techniques are divided into many types according to data types and domains.

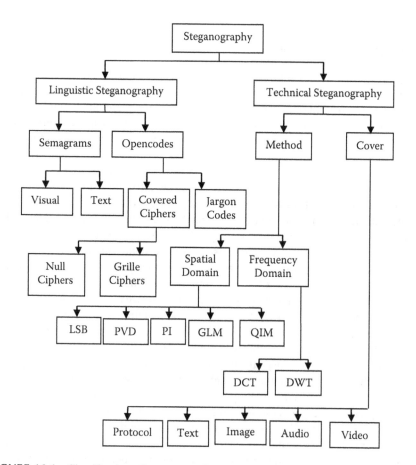

FIGURE 16.1 Classification of steganography.

Alternatively, steganography can be separated into reversible and irreversible information concealing strategies whether the cover article can be recuperated, where the data of mystery information can precisely be extricated. In irreversible information stowing away, the most uncritical pieces (LSB), substitution, and pixel-esteem differencing (PVD) strategies are utilized in conventional methods. In reversible information covering up, distinction extension (DE), histogram moving (HS), and expectation mistake development (PEE) are well-known in the spatial area [1,2]. Reversible information concealing procedures are utilized to address the issue of lossless implanting in touchy pictures e.g. military pictures, clinical pictures, and craftsmanship safeguarding [13,14].

16.3 STEGANALYSIS

Steganalysis is to recognize confidential information, decide shrouded information, and recuperate the concealed information. This can be classified into four: visual, primary, factual, and learning steganalysis.

The first type of steganalysis is to examine visual curios in the stego-pictures, where an attempt to get visual distinction by investigating stego-pictures is made. Primary steganalysis detected with the help of suspicious signs in the media design portrayal since the configuration is regularly changed when the mystery message is inserted. RS examination and pair investigation are remembered for the underlying steganalysis. Measurable steganalysis uses factual models to identify steganography strategies. Factual steganalysis can be isolated into explicit measurable and all-inclusive measurable steganalysis. Learning steganalysis likewise called daze steganalysis is one of the all-inclusive measurable steganalysis since cover pictures and stego-pictures are utilized as preparing datasets.

Other characterization of steganalysis can be partitioned into six classifications as appeared in Figure 16.2 [5]. It is relying upon what sort of assaults scientific inspector employments.

In Figure 16.2, all-inclusive or visually impaired steganalysis strategies depend on recognizing the mystery messages regardless of steganography methods. Contrasting and other steganalysis methods, general steganalysis procedure is exceptionally hard to track down extraction highlights, where AI strategies are frequently used to assemble, train, and assess models.

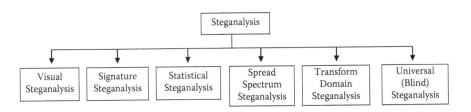

FIGURE 16.2   Steganalysis.

16.4 MACHINE LEARNING

AI is one of the manmade brainpower inventions, which is a PC-based technique for learning similar to the capacity of human cerebrum [6–9]. AI can be segmented into three primary classes: directed learning, solo learning, and support learning. Regulated learning is to get familiar with a capacity that maps a contribution to a yield by giving information yield sets, which are utilized regularly in discourse acknowledgment, spam location, and article acknowledgment. Unaided learning is the errand of gaining from test information that has not been marked, arranged, or classified. The primary use of solo learning is in the field of group examination, head segment investigation, vector quantization, and self-association. Fortification learning is worried about how to make moves to augment some thought of combined prize. Fortification learning is utilized in advanced mechanics, venture choices, and stock administration to learn activities to be performed. Deep Learning is a piece of AI that depends on learning information portrayals [14,19–27].

To create AI calculations, numerous structures are utilized e.g. Tensor Flow, Theano, Keras, Caffe, Torch, Deep Learning 4j, MxNet, CNTK, Lasagne, BigDL, etc. The accompanying area clarifies some of the systems in current.

In this chapter, many libraries are clarified, since scikit-learn, Tensor Flow, and Keras were utilized to test stego-pictures.

a. NumPy

NumPy is a notable and broadly used cluster-preparing bundle. A broad assortment of high multifaceted nature with numerical capacities makes NumPy incredible to handle huge multi-dimensional clusters and frameworks. NumPy is exceptionally helpful in dealing with straight variable-based math, Fourier changes, and irregular numbers. With the help of NumPy, we can characterize discretionary information types and effectively coordinate with most information bases.

b. SciPy

With rapid developments in AI, numerous Python engineers were developing python libraries for AI, particularly for logical and insightful registering. Travis Oliphant, Eric Jones, and Pearu Peterson, in 2001, chose to combine the majority of these pieces with a lot of codes and normalize it. The subsequent library was then named as SciPy library.

The current improvement of the SciPy library is upheld and supported by an open network of designers and conveyed under the free BSD permit. The SciPy library offers modules for direct variable-based math, picture advancement, joining introduction, exceptional capacities, Fast Fourier change, sign and picture preparing, Ordinary Differential Equation (ODE) settling, and other computational assignments in science and examination.

The basic information structure utilized by SciPy is a multi-dimensional exhibit provided by the NumPy module. SciPy relies upon NumPy for exhibit control subroutines. The SciPy library was worked upon to work with NumPy clusters alongside giving easy to use and productive mathematical capacities.

c. **Scikit-learn**

In 2007, David Cournapeau built up the Scikit-learn library as a component of the Google Summer of Code project. In 2010, INRIA included and accomplished public delivery in January 2010.

Scikit-learn [10] has a wide scope of directed and solo learning calculations that deal with a reliable interface in Python. The library can likewise be utilized for information mining and information examination. The primary AI works that the Scikit-learn library can deal with are characterization, relapse, grouping, dimensionality decrease, model choice, and preprocessing.

d. **Theano**

Theano is a python AI library that can go about as an improving compiler for assessing and controlling numerical articulations and lattice counts. Based on NumPy, Theano shows a tight mix with NumPy and has fundamentally the same interface. Theano can deal with Graphics Processing Unit (GPU) and CPU.

Usage of GPU design yields quicker outcomes. Theano can perform information serious calculations up to 140x quicker using GPU than a CPU. Theano can consequently evade mistakes and bugs when managing logarithmic and outstanding capacities. Theano has inherent instruments for unit testing and approval, thereby maintaining a strategic distance from bugs and issues.

e. **TensorFlow**

TensorFlow was produced for Google's inward use by the Google Brain group. TensorFlow is a well-known computational structure for making AI models [11]. TensorFlow is a wide range of toolboxes for building models at different degrees of deliberation.

TensorFlow uncovered entirely steady Python and C++ APIs. It can uncover, in reverse viable APIs for different dialects as well, yet they may be temperamental. TensorFlow has adaptable engineering with which it can run on an assortment of computational stages CPUs, GPUs, and TPUs. TPU represents the Tensor preparing unit, an equipment chip worked around Tensor Flow for AI and manmade brainpower.

f. **Keras**

Keras has more than 200,000 clients as of November 2017. Keras is an open-source library utilized for neural organizations and AI [12]. Keras likewise can run proficiently on CPU and GPU.

Keras, additionally, has a lot of highlights to chip away at pictures and text pictures that become convenient when composing Deep Neural Network code.

In addition to the standard neural organization, Keras upholds convolutional and repetitive neural organizations.

g. PyTorch

PyTorch has a range of devices and libraries that help PC vision, AI, and regular language handling. The PyTorch library depends on the Torch library and is open-source. The PyTorch library is simple for learning and utilizing.

PyTorch can easily be integrated with Python information science stack, including NumPy. You will scarcely have an effect among NumPy and PyTorch. PyTorch additionally permits designers to perform calculations on Tensors. PyTorch has a strong structure to assemble computational diagrams in a hurry and even change them in runtime. Different favorable circumstances of PyTorch incorporate multi GPU uphold, rearranged preprocessors, and custom information loaders.

h. Pandas

This is the most well-known Python library utilized for information examination with help for quick, adaptable, and communicative information that intended to chip away at both "social" or "named" information. This is an unavoidable library for settling pragmatic, genuine information examination in Python. Pandas are exceptionally steady, giving profoundly improved execution. Series (1-dimensional) and Data Frame (2-dimensional) are the two fundamental kinds of information structures utilized by pandas.

These two tools can deal with a large share of information prerequisites and use cases from most areas like science, insights, social, accounts, and, obviously, examination and different zones of designing.

Pandas upholds and performs well with various types of information including the following:

- Tabular information with sections of mixed information. For example, consider the information coming from the SQL table or Excel accounting page.
- Structured and unstructured time arrangement information. The recurrence of time arrangement need not be fixed, not normal for different libraries and devices. Pandas is incredibly quick in dealing with lopsided time-arrangement information.
- Other types of factual informational collections. The information is not necessarily named by any means. The Pandas information structure can handle it even without marking.

i. Matplotlib

Matplotlib library is utilized for an information representation with the 2D plotting to create distribution worth picture plots and facts in an assortment of organizations. The library assists with creating plots, mistake diagrams, and histogram disperse plots, bar graphs with only a couple lines of code. It gives an interface like MATLAB and is incredibly easy to use. It works with the help of GTK+, wxPython, Tkinter, or Qt and by utilizing standard GUI toolboxes to give an item arranged API that encourages developers to install charts.

16.5 STEGANALYSIS TESTS

a. Steganalysis Flowchart

Figure 16.3 shows the flowchart for steganalysis that utilizes machine learning algorithms.

Collection of information and testing of information requires data to be gathered and standardized to build precision. Specifically, vectorization should be done via element detection from pictures since pictures are utilized in steganography and steganalysis as a rule. In model structure, preparing, and assessing steps, different calculations can be used relying upon AI system.

b. Preparing Data

For the straightforward test, cover pictures and stego-pictures were readied that inserted the mystery information with 3-digit least huge pieces substitution. The mystery information was created by irregular capacity. Datasets can be set up by consolidating pictures and the mystery information differently.

c. Training Model

Keras and TensorFlow were utilized to build models, prepare models, and assess the yields. In this section, steganalysis and steganography of computerized pictures are basically treated as arrangement issues to make progressed AI (ML) strategies material. Three themes are covered: (1) compositional plan of convolutional neural organizations (CNNs) for steganalysis; (2) plan of measurable highlights for camera model grouping; and (3) certifiable altering location and confinement.

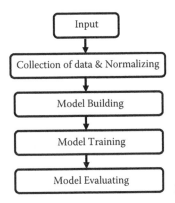

FIGURE 16.3 Steganalysis using machine learning.

16.6 CONCLUSION

In this chapter, different machine learning structures have been investigated to show the likelihood of steganography examination. We have discussed many libraries, which can be utilized for hiding secret data in the cover image. And a secured stego image can be obtained using machine learning algorithms. These machine learning algorithms can also be used for detecting secret data from the cover image i.e. the steganalysis process can be accomplished with the help of various libraries discussed in the chapter.

REFERENCES

1. Khan, A., Siddiqa, A., Munib, S., and Malik, S. A. 2014. A recent survey of reversible watermarking techniques, *Information Sciences* 279 (2014), 251–272.
2. Subhedar, M. S. and Mankar, V. H. 2014. Current status and key issues in image steganography: a survey, *Computer Science Review* 13 (2014), 95–113.
3. Nissar, A., and Mir, A. H. 2010. Classification of steganalysis techniques: a study, *Digital Signal Processing* 20 (2010), 1758–1770.
4. Cho, S., Cha, B. H., Gawecki, M., and Kuo, C. C. 2013. Block-based image steganalysis: algorithm and performance evaluation, *Journal of Visual Communication and Image Representation* 24, 846–856.
5. Karampidis, K., Kavallieratour, E., and Papadourakis, G. 2018. A review of image steganalysis techniques for digital forensics. *Journal of Information Security and Applications* 40, 217–235.
6. Musumeci, F. et al. 2018. An overview on application of machine learning techniques in optical networks. *Computer Science, Cornell University Library* (Oct. 2018), 1–27. https://arxiv.org/abs/1803.07976
7. Lee, J. H., Shin, J., and Realff, M. J. 2018. Machine Learning: Overview of the recent progresses and implications for the process systems engineering field. *Computer and Chemical Engineering* 114 (Oct. 2017), 111–121.
8. Schmidhuber, J. 2015. Deep learning in neural networks: An overview. *Neural Networks* 61 (Oct. 2014), 85–117.
9. Khan, M. A., and Salah, K. 2018. IoT security: Review, blockchain solutions, and open challenges. *Future Generation Computer Systems*, 82, 395–411.
10. Rana, A. K., Krishna, R., Dhwan, S., Sharma, S. and Gupta, R. 2019, October. Review on artificial intelligence with internet of things-problems, challenges and opportunities. In *2019 2nd International Conference on Power Energy, Environment and Intelligent Control (PEEIC)* (pp. 383–387). IEEE.
11. Rana, A. K. and Sharma, S. 2021. Contiki Cooja Security Solution (CCSS) with IPv6 routing protocol for low-power and lossy networks (RPL) in internet of things applications. In *Mobile Radio Communications and 5G Networks* (pp. 251–259). Springer, Singapore.
12. Rana, A. K. and Sharma, S. 2019. Enhanced Energy-Efficient Heterogeneous Routing Protocols in WSNs for IoT Application.
13. Ahmed, E., Islam, A., Ashraf, M., Chowdhury, A. I. and Rahman, M. M., Internet of Things (IoT): Vulnerabilities, Security Concerns and Things to Consider.
14. Veeramanickam, M. R. M. and Mohanapriya, M. 2016. IoT enabled futurus smart campus with effective e-learning: i-campus. *GSTF journal of Engineering Technology (JET)*, 3 (4), 8–87.
15. Cho, S. P. and Kim, J. G. 2016. E-learning based on internet of things. *Advanced Science Letters*, 22 (11), 3294–3298.

16. Kumar, A. and Sharma, S. 2020. Demur and routing protocols with application in underwater wireless sensor networks for smart city. In *Energy-Efficient Underwater Wireless Communications and Networking* (pp. 262–278). IGI Global.

17. Abbasy, M. B. and Quesada, E. V. 2017. Predictable influence of IoT (Internet of Things) in the higher education. *International Journal of Information and Education Technology*, 7(12), 914–920.

18. Kumar, A., Salau, A. O., Gupta, S., and Paliwal, K. 2019. Recent trends in iot and its requisition with IoT built engineering: A review. *Lecture Notes in Electrical Engineering*, vol. 526. Springer, Singapore. DOI: 10.1007/978-981-13-2553-3_2

19. Charmonman, S., Mongkhonvanit, P., Dieu, V., and Linden, N. 2015. Applications of internet of things in e-learning. *International Journal of the Computer, the Internet and Management*, 23 (3), 1–4.

20. Vharkute, M., and Wagh, S. 2015, April. An architectural approach of internet of things in E-Learning. In *2015 International Conference on Communications and Signal Processing (ICCSP)* (pp. 1773–1776). IEEE.

21. Dhawan, S., and Gupta, R. 2019. Comparative analysis of domains of technical steganographic techniques. 6th International Conference on Computing for Sustainable Global Development (INDIACom), March 2019.

22. Dhawan, S., and Gupta, R. 2020. Analysis of various data security techniques of steganography: A survey. *Information Security Journal: A Global Perspective*, 30 (02), 1–25.

23. Rana, A. K. and Sharma, S. 2020. Industry 4.0 manufacturing based on IoT, cloud computing, and Big Data: Manufacturing purpose scenario. In *Advances in Communication and Computational Technology* (pp. 1109–1119). Springer, Singapore.

24. Wang, Q., Zhu, X., Ni, Y., Gu, L., and Zhu, H. 2020. Blockchain for the IoT and industrial IoT: A review. *Internet of Things*, 10, 100081.

25. Rana, A. K., Salau, A., Gupta, S., and Arora, S. 2018. A survey of machine learning methods for IoT and their future applications.

26. Kumar, K., Gupta, E. S., and Rana, E. A. K. 2018. Wireless sensor networks: A review on "Challenges and Opportunities for the Future World-LTE". *Amity Journal of Computational Sciences (AJCS)*, 1 (2), 30–34.

27. Sachdev, R. 2020, April. Towards security and privacy for Edge AI in IoT/IoE based digital marketing environments. In *2020 Fifth International Conference on Fog and Mobile Edge Computing (FMEC)* (pp. 341–346). IEEE.

17 Gender Detection Based on Machine Learning Using Convolutional Neural Networks

A. Jaya Lakshmi[1], A. Rajesh[2], K. Aishwarya[2], R. Shashank Dinakar[2], and A. Mallaiah[3]

[1]Assistant Professor, ECE Department, Vardhaman College of Engineering, Hyderabad, India

[2]ECE Department, Vardhaman College of Engineering, Hyderabad, India

[3]Associate Professor, ECE department, Gudlavallereu Engineering College, Gudlavalleru, India

17.1 INTRODUCTION

Currently, the global population is increasing considerably. There are many factors in which one can face a problem when they travel with other people. All this makes our journey so difficult. Hence, introducing a system where there will be a safe environment for people to travel makes for a friendly environment. People following safety rules and regulations make good impression among everyone. Therefore, systems that monitor continuously and detect human faces will solve many problems in society.

Women's Public Transport Safety is one of the world's biggest challenges. Diverse steps are being undertaken by the authorities to make public transport safe and convenient for women. It is very important to note that men and women have very different travel needs. In developed countries, due to the lack of safe transportation options, many women have been forced to stay at home. Women are more likely to drive shorter distances and stop during the journey more often than men.

The establishment of a multi-disciplinary team is necessary if gender is to be effectively tackled in transportation projects. So, developing a module that identifies men and women in a particular place and their count can help the authorities to provide better needs to citizens and monitor them in a correct way.

In the field of protection, social services, social media, and surveillance, gender detection has many apps. There are many methods and designs to detect gender using facial photographs. Gender identification by facial images involves storage and verification of data by a process called Computer Vision & Machine Learning

DOI: 10.1201/9781003140443-17

using a Convolutional Neural Network (CNN) in datasets. Images are processed and tested by matching with datasets and implemented using simple hardware on Raspberry Pi programmed applying Java.

The proposed method detects genders and provides an output count of men and women from a group of personalities. This system produces an output using a live streaming method on Raspberry Pi programmed using Java. The entire streaming is videotaped and processed through testing datasets and output is displayed within a short span. The application of this model is majorly in public transportation systems and heavily crowded places.

17.1.1 PROBLEM STATEMENT

Automatic classification of gender, particularly with the rise of social networks and social media, has become a necessity due to an increasing number of applications. The proposed model outlines gender recognition related to computer vision & machine learning method by the CNN, which is used to eliminate various facial features. The facial data are then verified and the best features are added that will be functional for the training and testing of the dataset. They make use of a CNN with the aim of providing excellent machine performance efficiency with a gender detection rate for each of the databases, which is used in this study. This entire framework is implemented by means of the undemanding and easy hardware implementation on Raspberry Pi developed using Python for live streaming.

The goal of the project is to develop a low-cost module for the detection of gender in various aspects to solve many crimes and to maintain public safety. Machine Learning-supervised, Image Recognition-Artificial representations of the face region, Deep Learning-Convolution Neural Network, and Deep Learning-TensorFlow are the technologies used in the course of designing the module. Supervised learning is a machine learning algorithm where, with the aid of a testing dataset that has input and output sets, the input remains mapped toward the output. For mathematical computation, dataflow scripting, and various machine learning implementations, TensorFlow, an open-source library, is used. Computations of TensorFlow are represented as tasteful graphs of dataflow. CNN has gained a significant reputation in image feature extraction as one of the most efficient algorithms.

This report starts with an overview of the project where the importance of gender detection using machine learning has been discussed. We analyze the impact of gender detection to solve the number of crimes and even prevent them before happening. The methodology is all about developing the module using various technologies.

17.2 LITERATURE SURVEY

In [1], the author offers an overview of the in-depth approaches suggested over the last 6 years. Facial analysis involves a number of particular issues, such as face detection, identification of persons, recognition of gender and ethnicity, just to mention the most common ones. Significant research efforts have been devoted to the difficult task of age estimation from faces in the past two decades.

In [2], this paper presents traditional unfiltered benchmark techniques that reveal their incompetence in dealing with broad grades of differences in those unrestricted images. We suggest an innovative end-to-end CNN methodology in this work to attain a vigorous age set and cataloguing of gender. Function extraction and classification themselves are used in the two-level CNN architecture. Extracts of the feature extraction correspond to age and gender, whereas the classification categorizes the face images into the accurate age group.

In [3], the "Mitulgiri H" technique, by Gauswami and Kiran R Trivedi "relies on gender detection using facial mapping techniques that have been verified across numerous demanding stages of facial datasets, giving tremendous performance and effectiveness in the designed module." Different work on facial gender recognition was performed, introducing specific findings with their success rate for different databases. In their study, they focused on closing the gap between Google Cloud Vision technology and the real-time gender detection system. CNN was used in conjunction with an API cloud to implement gender detection to accurately predict the data class (man or woman) on an appropriate inexpensive, credit-card-sized Raspberry Pi board furnished with a camera section. In [4], scanning of a human face, weight, gender, and age using various algorithms has been put forth. The paper proposed about finding the best techniques to detect gender, weight, and gender efiiciently. The designed module was cost-effective and a credit-card-sized mini-computer. A cost-sensitive ranking algorithm was used to detect human age based on face pictures. Two components called facial extraction and estimator learning were implemented for an effective age estimator. Face detection was achieved using the Jones viola algorithm, which uses a Haar feature-based cascade. The local binary pattern approach most effective for face recognition was used for age estimation feature extraction. Face and mouth movements were processed with input datasets using CNN. Lastly, weight was divided into three parts: underweight, average weight, and overweight. Python coding detected all these functions. In [5], the author depicts approaches for gender detection and explained gender recognition using facial images with various preprocessing processes. Processing methods were proposed using ellipse face pictures, Gabor philtres, Adabost learning and SVM classifier. In [6], Joseph Lemley et al. stated that Artificial Intelligence was used in a high range for gender detection. The paper was all about justifying the techniques and classifying them. In [7], the examination detailed about facing the challenges in performing gender detection by means of a CNN. After training CNN, a well-trained and vigorous system that can categorize faces of men and women have been generated. In [8], 8the author addressed the problem of a multi-tasking algorithm using a simple CNN. Task-based regularization and domain-based regularization were two methods used to calculate the parameters required for pose and smile estimation. Later, network architecture was implemented to combine all these features into a single algorithm. Using CNN output was evaluated. In [9], Jos van de Wolfshaar et al., by refinement, a pre-trained neural network, the applicability of deep convolutionary neural networks to gender classification has been investigated. Deep neural networks were used to support the vector machines for gender detection. In [10], the authors proposed a strategy that relies on a deep CNN-based approach for tough facial gender and smile classification. To improve predictive performance, multi-view learning and also the strategic-to-specific perfect-tuning method were proposed. A task-aware cropping scheme is proposed to further enhance the model's performance. In [11], Linnan Zhu et al. described how gender detection can be performed in different ways. This architecture completed two tasks, detection of age and gender, in a cost-effective and efficient way. This model achieved competitive output with modern methods as well as is often much quicker than other modules. In [12], the face and gender can be detected using two different datasets but within the same neural network. The face detector in this system had an average detection accuracy of 97.9%. Face Recognition and Gender Classification Modules used a neural network architecture evaluated on two separate datasets i.e. Database of the Site and Bio ID. In [8], the authors presented a multi-purpose algorithm using a single deep CNN for simulated face identification. A multi-task learning system is used that regularizes CNN's shared parameters. In [13], emerging computer vision and pattern recognition applications in mobile devices and networked computing involve the improvement of resource-limited procedures. In this sense, linear classification techniques have a significant part to play. The gender classifiers of the SVM are superior to the rest when appropriate training statistics and computer resources are obtainable.

17.3 IMPLEMENTATION AND DESIGN

17.3.1 HARDWARE ARCHITECTURE AND BLOCK DIAGRAM

The hardware architecture describes the components used to build the entire module. It also helps in evolving the features of the module and its operations.

Figure 17.1 shows a block illustration of the flow of the project where different procedures are used to get the estimated output.

17.3.2 HARDWARE TOOLS

We use the Raspberry Pi board with 4 GB of onboard storage in this real-time gender classification device module, which is interfaced with a camera module and a display screen to extract input images using a computer vision library. Later, by comparing inputs with stored datasets, the code continues to run and later generates output in the result segment. It will indicate the overall count of men and women. The hardware tools used for building the gender detection module using CNN are shown below in Figure 17.2.

1. Raspberry Pi 4 board
2. USB camera
3. 5V power supply
4. Display
5. HDMA converter
6. 4 GB SD card

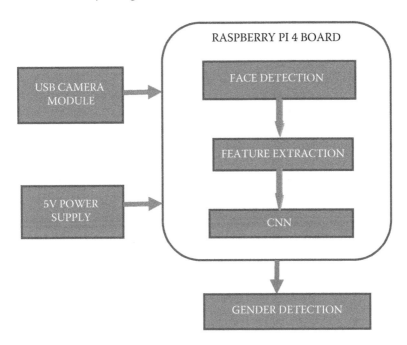

FIGURE 17.1 Gender detection block diagram.

FIGURE 17.2 Raspberry Pi 4 board.

17.3.2.1   Raspberry Pi 4 Board

Features:

A low-cost, credit-card-sized computer that uses a regular keyboard and mouse and plugs into a monitor or television device is the Raspberry Pi shown in Figure 17.2. It is a lightweight, accomplished machine that encourages individuals of all ages to discover programming plus study how to programme in languages such as Scratch as well as Python. It can do all you would suppose a desktop computer to prepare, after surfing the internet to viewing high-definition films, to constructing spreadsheets, word-processing, then playing sports.

In addition, the Raspberry Pi partakes the capability to communicate by the outer world, besides it has been included in a vast range of immersive developer projects, including weather stations and infra-red cameras tweeting birdhouses with music machines and parent detectors. We like to see where the Raspberry Pi is used to understand how machines work for kids around the world. There are actually four replicas of the Raspberry Pi. They are Model A, Model B, Model B+, and the Measurement Module. Both versions have the same BCM2836 CPU, but some characteristics of the hardware vary.

The Model B+:

The Raspberry Pi 3 Model B+ is the newest product in the Raspberry Pi 3 series, containing an updated 64-bit quad-core 1.4GHz processor using a built-in metallic heat sink, dual-band 2.4GHz, and 5GHz wireless LAN, faster (300 mbps) ethernet, and PoE capability through a separate PoE HAT.

The Raspberry Pi 3 Model B+ keeps the same mechanical impression equally both the Raspberry Pi 2 Model B and the Raspberry Pi 3 Model B. Adafruit made/ branded circumstances will quiet suit, on the other hand some other cases might not, and particularly those rely continuously on the position of the part or remain installed in the heatsink (Table 17.1).

17.3.3   Software Architecture and Algorithm

Step 1: Save the entire code in the Raspberry Pi board in its 4 GB memory card.

Step 2: Install it into the Raspberry Pi board as soon as the whole programme is loaded into the memory card.

Step 3: Now link the 5V power supply to the Raspberry Pi board through a cable.

Step 4: Link the display to the Raspberry Pi board by using the HDMA converter.

Step 5: Connect a USB camera, mouse, and keyboard to the Raspberry Pi through its ports.

Step 6: As soon as the connections are properly done ON the power supply of the Raspberry Pi board and monitor.

Step 7: Now, by opening the editor icon an instruction window will be opened.

TABLE 17.1

Specifications of Raspberry Pi 4 Board

CPU type/Speed	700 MHz Low Power ARM1176JZFS Applications Processor
# of CPU Cores	4
RAM Size	1,2,4 GB LPDDR4
Integrated Wi-Fi	2.4GHz and 5GHz
Gigabit Ethernet	Yes
Bluetooth	5.0 BLE
USB	2 USB 3.0,2 USB 2.0
HDMI	2μ, 4k video
Video Decode	H.265(4kp60), H.264(1080p30)
Video Encode	H.264(1080p30)
Open GL ES	1.1,2.0,3.0 graphics
Microprocessor	Broadcom BCM2837 64bit Quad Core Processor
Processing voltage	5V
Chip Broadcom	BCM2836 SOC
Core architecture	ARM11
Memory	512MB SDRAM
Dimensions	85 x 56 x 17mm

Step 8: Giving instruction in the editor window.
Step 9: Click cd age_gender_python/ andclickENTER.
Step 10: Now click sudo python3 agegenderpredict.py and click ENTER.
Step 11: As soon as we click on ENTER result window opens.
Step 12: The webcam takes input images and if face detected gender will be displayed in the terminal.
Step 13: Otherwise, the program stops execution.

17.3.3.1 Installation Steps on Raspberry Pi Board

Step 1: Developing the program using python

The requirement program code is developed as it takes input and stores temporarily. After that, the input compared with a module named weights is shown below in Figure 17.3.

Step 2: Dumping code into Raspberry Pi

By taking SD card, using Raspbian OS we had installed program into the card. After that, the card is inserted into the Raspberry Pi as shown below in Figures 17.4 and 17.5.

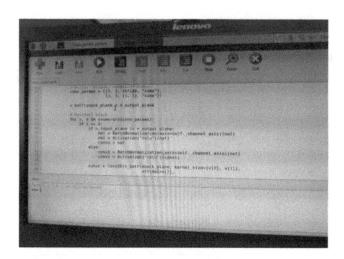

FIGURE 17.3 Technologically advanced code using python.

FIGURE 17.4 Installing Raspbian OS.

Step 3: Connecting external devices to Raspberry Pi

After the installation process now connecting USB camera, keyboard, and mouse are shown in Figure 17.6.

Step 4: Now connecting display to Raspberry Pi

By using an HDMI converter, the display is connected to Raspberry Pi shown in Figure 17.7.

FIGURE 17.5 Inserting SD card into Raspberry Pi.

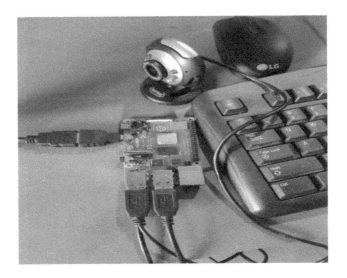

FIGURE 17.6 Connecting external devices to Raspberry Pi.

Step 5: Now connecting 5V power supply to the Raspberry Pi shown in above Figure 17.8.

Step 6: On the monitor

When the monitor is ON, the screen has a terminal shown

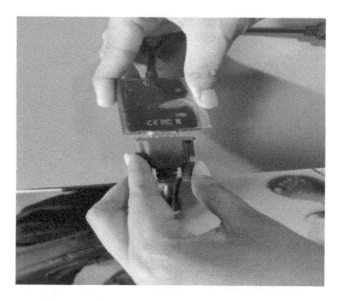

FIGURE 17.7 HDMI assembly.

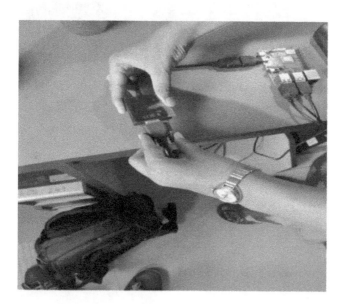

FIGURE 17.8 Power supply association.

Step 7: Open the terminal and enter the file details

1. cd age_gender_python/
2. Click enter
3. sudo python3 agegenderpredict.py
4. Click enter

FIGURE 17.9   Face recognition with blue Tag (Female: 0 Male: 1).

Step 8: Now the result window opens.

The webcam starts taking the input image and if a face is detected a blue quadrilateral box will appear around the face (Figure 17.9).

Step 9: Output

When the face is detected, the image within the blue box is processed and compared with the trained dataset. Then, the count along with gender will be displayed (Figure 17.10).

17.4   SIMULATION RESULTS

All the hardware components are connected as per the block diagram and the power source is specified to the Raspberry Pi boarding. The webcam starts capturing the images and forwards them to the processing unit presented. Now, the HDMI cable of the monitor remains connected to the Raspberry Pi board. When the power supply is on, the monitor turns ON and the Rasbian operating system is displayed. As soon the monitor gets ON the instructions to run the program are entered in the terminal box. After giving all the instructions, the terminal runs the backend code and starts running the program.

After the instructions are given the program starts executing the output.

As soon as the program starts executing a result box will be opened and the webcam starts capturing the images displayed. Webcam captures images and they are processed. If a human face is detected the blue rectangular box will be drawn around the face. It nearly takes 30 to 40 seconds delay to generate the output and then the gender will be displayed in the result box. As in the above figure, only one male face is captured so the result box shows the count as males: 1 and females: 0. Even in the result window, the gender is displayed as M for male and F for females as shown in Figure 17.11.

FIGURE 17.10 Face detected the Image with blue Tag (Female: 1 Male: 2).

FIGURE 17.11 Single gender detection and count display on monitor.

FIGURE 17.12   Dual gender detection and count display on monitor.

Now, when the program is executed again with two males before the webcam their faces are processed and compared with the facial images stored. After comparison, the blue boxes will have appeared around the detected human faces and their gender will be displayed in the result box. In the above figure as there are two males, the output will be shown after 30 sec delay as males: 2 and females: 0 shown in above Figure 17.12.

As the project is all about multiple face detections, when three human faces of different genders are processed the output will be displayed as count of males: 2 and females: 1 shown in above Figure 17.13. In this way, multiple gender detections are done with less delay and more efficiency.

17.5   CONCLUSION

An automated method of gender recognition focused on convolutional neural networks (CNN) was developed by the designed module. We assume that not only a single facial appearance but also multiple facial detections can be achieved with improved efficiency using CNN techniques by looking at the results of our project. Our present multimodal solution can be improved in multiple ways. The system is made up of a face detector and a gender classifier that compares the stored datasets with the input images. With a delay of 30 seconds, the face recognition system analyses the whole input image and produces the output. Not only obtaining competitive performance, but our proposed approach also runs much faster and with greater efficiency. The findings reveal

FIGURE 17.13 Multiple gender detection and count display on monitor.

that the CNN is more robust and also more stable instrument than general classifiers, such as SVM and Random Forest.

REFERENCES

1. Carletti, A. S. (2019). "Age from Faces in the Deep Learning Revolution." *IEEE Transactions on Pattern Analysis and Machine Intelligence,* vol. 3, pp. 1.
2. Viriri, O. A.-A. (2020) "Deeply Learned Classifiers for Age and Gender Predictions of Unfiltered Faces." *Hindawi, The Scientific World Journal,* vol. 2020, Article ID 1289408, pp. 12.
3. Trivedi, M. H. (2018). "Implementation of Machine Learning for Gender Detection using CNN on Raspberry Pi Platform." *IEEE Proceedings of the Second International Conference on Inventive Systems and Control,* pp. 603–613.
4. Zamwar, S. C., Ladhake, S. A., & Ghate, U. S. (2016). "Human Face Detection and Tracking for Age Rank, Weight and Gender Estimation Based on Face Images Using Raspberry Pi Processor." *International Journal of Engineering Research and Application,* vol. 5, pp. 16–21.
5. Huchuan Lu, H. L. (2007). "Gender Recognition Using Adaboosted Feature." *Third International Conference on Natural Computation (ICNC).*
6. JosephLemley, S. D. (2016). "Comparison of Recent Machine Learning Techniques for Gender Recognition from Facial Images." *MAICS Conference,* pp. 97–102.

7. LovekeshVig, A. A. (2014). "Convolutional Neural Networks to discover cognitively validated features for Gender Classification." *IEEE International Conference on Soft Computing & Machine Intelligence*, pp. 33–37.
8. Rajeev Ranjan, S. S. (2017). "An All-In-One Convolutional Neural Network for Face Analysis." *IEEE 12th International Conference on Automatic Face & Gesture Recognition*, pp. 17–23.
9. Jos van de Wolfshaar, M. F. (2015). "Deep Convolutional Neural Networks and Support Vector Machines for Gender Recognition." *IEEE Symposium Series on Computational Intelligence*, pp. 188–195.
10. Kaipeng Zhang, L. L. (2016). "Gender and Smile Classification Using Deep Convolutional Neural Networks." *Proceedings of the IEEE Conference on Computer Vision and Pattern Recognition (CVPR) Workshops*, pp. 34–38.
11. Linnan Zhu, K. W. (2016). "Learning a Light Weight Deep Convolutional Network for Joint Ageand Gender Recognition," *IEEE 23rd International Conference on Pattern Recognition (ICPR)*, pp. 3282–3287.
12. Tivive, F., & Bouzerdoum, A. (2006). "A Gender Recognition System Using Shunting Inhibitory Convolution Neural Networks." *2006 IEEE International Joint Conference on Neural Network Proceedings*, pp.5336–5341.
13. Juan Bekios-Calfa, J. (2011). "Revisiting Linear Discriminant Techniques in Gender Recognition." *IEEE Transactions on Pattern Learning and Machine Intelligence*, vol. 2, pp. 858–864.

18 Smart Technologies and Social Impact

An Indian Perspective of Contactless Technologies for Pandemic

Rashmi Bhardwaj[1], Varsha Duhoon[2], and Mohammad Ayoub Khan[3]

[1]Professor of Mathematics, University School of Basic & Applied Sciences (USBAS), Head, Non-Linear Dynamics Research Lab, Guru Gobind Singh Indraprastha University, Delhi, India
[2]USBAS, GGS Indraprastha University, Delhi, India
[3]College of Computing and Information Technology, University of Bisha, Bisha, Saudi Arabia

18.1 INTRODUCTION

This is not the first time in the history of mankind that the world is affected by a pandemic. There have been a number of diseases and infections, which affected different parts of the globe or many countries in the past, and here again, a new virus outbreak has made the situation worse for our survival. A pandemic or outbreak can be defined as the sudden beginning or spread of an infection or virus that can cause destruction on a large scale. Different types of viruses, infections, and diseases have come about and on the basis of their spread and life span we can study the outbreak of diseases, which are classified as Epidemic, Endemic, and Pandemic. An epidemic can be defined as the spread of a kind of disease at a very high rate into a large population in a time period which is not long, and so the development of a privacy-preserving Bluetooth protocol to support Contact Tracing has been done to trace contact and to make people aware of their exposure [1]. The WHO report shows the outbreak of SARS in multiple countries and how it affected people across the globe and caused death, and different methods adopted for prediction are also discussed [2]. A smart electronic solution was developed and discussed, which will help in reducing the new cases of COVID-19 [3]. Such diseases spreading at a rapid rate affect the lives of many in a particular area where the disease-causing virus or any other agent has spread. For example, the meningococcal infection, an infection

DOI: 10.1201/9781003140443-18

that had spread at an alarming rate of 15 cases per 100,000 people for two continuous weeks, was an epidemic [4]. The Virgin soil epidemic devastated the native population all across the globe in the wake of European conquest. A pandemic can be defined as the spread of infection or disease over a large area or more than one continent or across the globe affecting a large number of people. A few among the pandemic diseases are smallpox and tuberculosis. In the past, the "Black Death," also called "the Plague," was a pandemic that had resulted in deaths on a large scale across the globe, the numbers are 75–200 million people in the fourteenth century [4]. Another pandemic in history was that of influenza in 1918. At the present time, the current pandemics are HIV/AIDS and COVID-19, the latter of which has already affected many countries across the globe with the number of cases rising to 84,67,178 worldwide, affecting 213 countries and territories [4]. The spread of COVID-19 to different parts of the globe, affecting regions such as North America, South America, Europe, and Asia, has resulted in 4,51,954 deaths, and the most affected countries thus far have been the USA, Brazil, Russia, India, the UK, Spain, Peru, Italy, Chile, Iran, and others [5].

An endemic can be defined as a type of disease that spreads by bacterial or viral agents and is always present in society. Malaria is an endemic in Africa, but due to knowledge, education, and awareness, the cases that have been reported are decreasing. Table 18.1 gives a list of a few of the most common infectious diseases in human beings [6]. In short, we can define endemic, epidemic, and pandemic as follows [7]:

- Endemic: Refers to a type of disease that is present in a place or population.
- Epidemic: Refers to an outbreak that has affected a large number of people at a time and spreads across one or many communities.
- Pandemic: The word pandemic is an explanation of the word epidemic when the spread is at a global level.

Now, an epidemic is declared a pandemic by the World Health Organization.

Currently, an example of a pandemic is COVID-19. On December 13, 2019, cases were reported with patients suffering from pneumonia due to unknown cause, in Wuhan city, in the province of Hubei in China, which was further reported to the WHO [5]. After some time, the unknown cause came to be known as a new virus in January 2020, and, hence, it was seen that within a few months the number of cases being reported was rising each hour and day, and the cases were not only reported in China but also across the globe, resulting in declaration of a pandemic on March 11, 2020; after few days, on March 15, 2020, more than 150,000 cases were reported across the globe with 123 countries affected and more than 5,000 deaths [5]. The spread of a disease or virus is declared to be a pandemic on the basis of 6 phases. The first phase is a low risk and the sixth phase is a complete pandemic. The phases are as follows [7]:

P1: Presence of the virus in animal(s), but not infectious to humans.
P2: A virus from animals affects humans.

TABLE 18.1
Diseases, Agents, Transmission [6]

Disease	Infectious Agent	Mode of Transmission
CHOLERA	BACTERIUM	CONTAMINATED FOOD/WATER
DIPHTHERIA	BACTERIUM	CONTAMINATED FOOD/WATER
HEPATITIS A	VIRUS	CONTAMINATED FOOD/WATER
HOOK WORM	PARASITIC WORM	CONTAMINATED FOOD/WATER
SALMONELLA	BACTERIUM	CONTAMINATED FOOD/WATER
TAPEWORM	PARASITIC WORM	CONTAMINATED FOOD/WATER
TYPHOID	BACTERIUM	CONTAMINATED FOOD/WATER
CHIKEN POX	VIRUS	DROPLET
COMMON COLD	VIRUS	DROPLET
INFLUENZA	VIRUS	DROPLET
MEASLES	VIRUS	DROPLET
MOMPS	VIRUS	DROPLET
POLIOMYELITIS RUBELLA	VIRUS	DROPLET
SCARLET FEVER	BACTERIUM	DROPLET
SMALL POX	VIRUS	DROPLET
TUBERCULOSIS	BACTERIUM	DROPLET
WHOOPING COUGH	VIRUS	DROPLET
HIV/AIDS	VIRUS	SEXUAL CONTACT, BLOOD
SYPHILIS, YAWS	TREPENOMAL BACTERIA	SEXUAL CONTACT
PLAGUE	BACTERIUM	FLEA & RAT VECTOR
TYPHUS	RICKETTSIA	FLEA & RAT VECTOR
ENCEPHALITIS	VIRUS	INSECT VECTOR
MALARIA	AMOBEA	INSECT VECTOR
ONCHOCERCIASIS	PARASITIC WORM	INSECT VECTOR

P3: Spreading of a disease or infection among humans, which also transfers from human to human, but does not have a community level outbreak.

P4: Transferring from human to human causes a community-level outbreak.

P5: The disease has now spread from one country to another and another, thus, affecting more than one country.

P6: Apart from P5, there is a community-level outbreak in more than one country.

By the time the disease or infection reaches the P6 stage, it is then termed a global pandemic [7]. Tracking and localization using IoT have been studied and proven to be quite effective. The new ways and fields are an adaptive method of IoT and prove the worth of the application of IoT in different fields and different sectors [8–13]. Research about IoT includes and connects technologies to gain information about different objects [14–16]. The use of RFID and IoT shows important problems and interoperability [17]. IoT is applied in smart cities and shows how it is beneficial for the society [18], [19]. IoT applications, technology, systems, applications, methods, and models are discussed [20]. A track paper system by the application of RFID, IoT, and cloud technologies in a pharmaceutical company is developed to keep a record of the waste formed and to record the functioning [21]. The application of IoT in food production, transportation, and distributing food in retail is studied [22]. IoT and other emerging technologies and their relationship with Big Data analytics, cloud, and fog computing are studied [23]. The major challenges related to the development of IoT are discussed [24]. IoT is studied as a whole and how it can be a semantic web [25]. The application and uses of RFID are discussed in detail [26]. A new design is formed, which proposes an International Telecommunication Union-based standardization (ITU-T) and discusses future directions in wearable IoT [27].

Author's Name	Study	Year
G. Yang et al.	discuss the application of fused IoT devices with in-home healthcare services	2014
F. Noonan et al.	discuss automation of paper tracking system by RFID, IoT methods	2018
R. P. Singh et al.	discuss IoT applications and how IoT helps in fighting against the situation of COVID-19	2020
Ting et al.	discuss the application of-IoT, big-data analytics	2020
Pan X.	deals with the application of AI which helps in prior tracking and tracing to fight COVID-19	2020
A. A. R. Alsaeedy et al.	discuss a new strategy for areas with high human density and mobility, which are at risk for spreading COVID-19	2020
N. Zheng et al.	includes the comparison of old and new hybrid AI model and proposed model reduce the error in forecasting	2020
M. Quayson et al.	discuss the advanced digital applications to help the small farmers.	2020
Q. Pham et al.	deals about the tracing using computer-based tools and analyzing the COVID-19 patients	2020
T. Tamai et al.	discuss the relationship between the contact limitation and tracing	2008
M. Amin et al.	application of IoT and security challenges of different factors & problems in COVID-19 time	2015
F. Wu et al.	discuss IoT and sensors analysis of the application of IoT for safety and health	2019
R. Khan et al.	discuss the application of smartphone technology for tracking and tracing the spread of a virus	2012

Author's Name	Study	Year
V. Chamola et al.	studies about COVID-19 and its impact on the health and economy of the country.	2020
A. A. Hussain et al.	discuss an application that will help doctors track and trace the COVID-19 affected persons and provide them treatment accordingly	2020
Z. Han	the proposed algorithm has an overall accuracy in the screening of COVID-19	2020
F. Rustam et al.	deals with the application of different methods among which ES performs best among other models	2020
M. ABasset et al.	proposed hybrid COVID-19 detection model based on IMPA for X-ray image division	2020
Bhardwaj R. et al.	discusses that smart technologies provide a better perspective for health improvement and ease of access but also creates a digital divide among people during this crisis time as smart technologies have a great impact on social behavior	2020

A system is proposed using an open-source controller and a GPS/GSM/GPRS module for the application of data transfer [28]. A practical study was conducted to understand fog computation, application, and analysis [29]. The application and effectiveness of the IoT philosophy to implement how to fight COVID-19 is discussed [30]. The reasons how all industries are affected by the spread of COVID-19 and how it has made everyone's lives miserable during this pandemic [31]. The application of IoT to provide information and monitoring systems during the COVID-19 epidemic is analyzed and this is expected to help in dealing with the challenges by an automated and transparent treatment process using IoT, which can be applied to trace and fight against COVID-19 [32]. The application of IoT, Big Data analytics, AI, and blockchain for traditional public-health strategies to fight against COVID-19 is discussed [33]. The application of AI to fight the COVID-19 pandemic is used to help in accurate screening, tracking, forecasting current and future patients, and in prior detection and diagnosis of infection [34].

A new design is introduced using *fog technology*, which will help in generating spot and fast analysis and help to take all required action in time [35]. The integration of AI with X-ray and CT is used to analyze the effects of the technology in the field of COVID-19 tracking [36]. The application of Machine Learning to detect COVID-19 cases with the help of a mobile phone-based web survey would help in the tracking of the spread of the virus during quarantine time [37,38]. A mechanistic model is proposed to track and trace any misinformations in the medical field and solve the problem faced on the social media platform [39]. The application is used for the classification of models to predict Twitter users who are more likely to use controversial terms and to help mitigate the impact of the COVID-19 outbreak [40]. A new strategy is identified for areas with high human density and mobility, which are at risk for spreading COVID-19 with special qualities for forecasting the reposted quantity of every kind of information [41]. The application of different methods among which exponential smoothing (ES) performs better as compared to support vector machine (SVM) [42]. A hybrid COVID-19 detection model is proposed based on the improved marine predator's algorithm (IMPA) for X-ray image segmentation [43]. The comparison of the traditional epidemic models and the proposed hybrid Artificial Intelligence (AI) model proves that the proposed model can significantly reduce the errors of forecasting results [44]. The application of AI and other image studying methods is analyzed to improve the image quality for better tracking and tracing of COVID-19 [45]. The application of digital technologies for farmers is to avoid major issues from damaging the livelihoods of society, which include post-COVID-19 digital inclusion in developing countries [46]. The high accuracy of computer-aided diagnostic tools for better speed and correct prediction and detection rate of COVID-19 diagnosis is suggested by an algorithm for screening COVID-19-affected patients [47]. The relationship between controlling the contact and tracking the contacts is discussed and results on its effectiveness are analyzed [48].

The application of IoT with its growth in India, which contains security issues, an analysis of the risk factor, and challenges in the Indian perspective are discussed [49]. Smart energy technology is applied to reduce the energy cost and environmental impact in building life cycles [50]. The application of sensors, which includes the analysis of the application of IoT for the safety and health of workers, is discussed [51]. The application of e-health architecture using IoT for data acquisition, fog for data pre-processing and short-term storage, cloud for data processing, are used to analyze long-term storage [52]. The application of fusion-based IoT devices with in-home healthcare help is discussed, which helps in the rise of improved experiences and accuracy[53]. Different kinds of IoT devices, which can be used to increase the efficiency of the health of people are analyzed and have no errors in the way of medication and medicine delivery [54,55]. IoT is becoming a major platform for many services and applications, and using Raspberry Pi not just as a sensor node but also a controller is utilized [56]. A model is made and used for medical persons who can further apply this algorithm for the study[57]. The application and use of technologies are helpful to reduce the visits of doctors because information and details about patients' health will be directly transferred to doctors' monitor-screens from any part of the country where patients live [58], and the experiences of medical practitioners in terms of using the Big Data of COVID-19 is used for further discussion [59,60]. The type of device that will help in the monitoring of the person using it to decrease the spread of Coronavirus (COVID-19) is discussed [61].

How can we control or prevent the pandemic?

It is extremely important that an epidemic should not reach the level of a pandemic. Hence, set rules or policies are made to control the spread at the initial stage. There are many steps that have been proven to help and will be effective in controlling and containment of the virus. The steps to control are as follows:

- Control
- Identify cases

- Trace contact
- Quarantine
- Isolate
- Protect

The world today is facing serious problems due to the spread of COVID-19, which has adversely affected the economy across the globe. In this difficult time, IoT-based devices to support the health sector have advanced quite rapidly. In a time where the world was looking forward to greater advancement in technology, with a crisis in line with nuclear war and changes in the climate across the globe leading to a rise in temperature and rise in the water level, no one could have imagined that a catastrophic disaster, such as COVID-19, would hit countries and affect the growth of technology and advancement in all the sectors. All healthcare workers are facing challenges to find a cure for the disease and also to find a vaccine to stop the spread of the COVID-19. However, in this hard time, the IoT has taken over in almost every field and especially in the healthcare sector by introducing and developing new applications that will help in reducing human contact and promote more dependency on the technology and, hence, will help in following social distancing protocols [62]. It is seen that there is a constant increase in people getting affected by COVID-19 across the globe. At the end of May 2020, the recorded cases were 57,92,992 across the globe and 3,57,480 deaths were reported. In such hard times, there is a need for technology that can help the health sector in different ways.

Digital Divide is another concept that means the gap in the area demographically in terms of who has access to smart technologies and who does not, or who has restricted access to using smart technology. Smart technology may include telephones, televisions, personal computers, and the Internet. The contribution of the digital technology divide in healthcare can be seen as people are using smartphones and, in countries like India, the majority of people are using different applications for tracing COVID-19 symptoms, contact tracing, and also to see the number of people affected in the area nearby and, hence, follow all safety measures. Yet, there are places where people do not have a proper Internet connection and no smartphones to help them access such applications, which are for their own benefit [63–65].

18.2   TAXONOMY OF TECHNOLOGIES

18.2.1   THE INTERNET OF THINGS

IoT refers to inter-connected computational devices, mechanical and digital machines that have unique identifiers (UIDs), and the capability of transferring information using a network without the movement of human-to-human and human-to-computer interaction. IoT refers to the network of objects connected to the Internet, which are capable of collecting and exchanging information. IoT has brought a revolutionary change in the way that devices were used and the way that they are used now. The technology has brought such advancement by helping in collecting the data that can further be used to upgrade user experiences. The application of IoT in different aspects is discussed [66]. The sensor networks based on classification and mitigation are discussed [67]. IoT in terms of innovation based on strategic sharing is analyzed [68]. Message Queuing Telemetry Transport (MQTT) and Common Offer Acceptance Portal (COAP) are discussed [69]. IoT, in terms of the recent advances and new solutions generated using IoT technology, is applied [70]. IoT and blockchain for implementation in smart contracts are analyzed [71]. IoT for future prospects is discussed in detail [72]. Future challenges and the application of IoT in different fields are discussed [73]. The Internet of Things (IoT), the Internet of Everything (IoE), Internet of Nano Things (IoNT), and the future trends of IoT are studied by different authors in different areas where data security and smart contracts are implemented in the field of healthcare, database management for the purpose of data sharing, smart environments, and monitoring systems for health [74–82]. The protection of data is extremely important, not just when the data is being collected, but also when the data is stored for a whole life. IoT has made devices smarter due to the connection of devices with the Internet, which has expanded the boundaries. IoT can be classified into the following categories:

- Collection of information and sending it, which includes sensors for temperature, motion, moisture, air quality, and light. The sensors are used in collecting information from the environment and help in clearly analyzing the extracted information.
- Things that gain information and further work on it, such as smart TVs and 3-D printers.
- Things that can do both, which includes an IoT-based agriculture system using the gathered information about soil moisture, temperature, and the amount of water required by crops to make decisions and solve problems.

The working of IoT depends on inputs in the system; hence, the protection of information becomes very important for its whole life and now is when blockchain is combined with IoT. IoT will help the healthcare sector by keeping track of the people affected by COVID-19, pre-screening, diagnosing, cleaning, disinfecting, using drones in innovative ways, and in-home infection reduction.

18.2.2 APPLICATION OF IoT IN DIFFERENT SCENARIOS OF COVID

18.2.2.1 Quarantine Tracking

It is important to keep a record of the spread of COVID-19 in people who are affected by it or at risk. However, it is easier said than done, in a world with huge populaton. Hence, the world is looking forward to IoT- and GPS-based applications to keep a record and, when required, restrict people's movements. Russia, Poland, Singapore, South Korea, and India are some countries that are using such apps for help. In Hong Kong, the quarantine procedure was implemented from the airport itself by giving them wristbands; also, they were given quick response (QR) to record their movements. This technology is geofencing, in which a virtual perimeter is created with the help of GPS, RFID, WI-FI, Bluetooth, and cellular network [41].

18.2.2.2   Pre-Screening

Hospitals and all medical centers started using telemedicine to diagnose and answer problems regarding COVID-19. In such a situation, software companies collaborated with hospitals and medical centers to set up chatbots on websites and mobile apps. The chatbots help by asking a series of questions to screen visitors according to the severity of their conditions. In this way, more and more time of doctors is diverted toward treating the patients and not answering repeated questions [83].

18.2.2.3   Disinfecting and Cleaning

COVID-19 virus spreads by contact or by touching any object that has come in contact with an infected person. Hence, proper sanitization, cleaning, and disinfecting measures need to be implemented in medical facilities. COVID-19 has shown how robots can be used to sanitize and clean healthcare centers and public places. They disinfect places by emitting high-intensity UV rays, which kill the virus by tearing apart its DNA. They are based on Wi-Fi and can be controlled using apps. In the present time, this technology is being used in China, Italy, and the USA.

18.2.2.4   Drone-Based Surveillance

Due to the spread of COVID-19, drones have found some innovative uses, including:

- To keep a check on and enforce the stay-at-home orders in Spain and China
- To disinfect the contaminated hotspot of Daegu, South Korea
- To provide medical samples and quarantine materials in China
- To check the temperature of people who are in quarantine with the help of infracted thermometers on drones as patients stand on their balconies

18.2.2.5   In-Home Infection Reduction

Coronavirus spreads by touching anything that has come in contact with someone else who is already affected by COVID. Hence, to maintain social distancing, it is important to avoid touching objects. IoT-based smart speakers, lights, and security systems can be used to open doors and switch on lights. IoT also helps in the ease of video conferencing and also helps in virtual meetings.

18.2.3   APPLICATIONS OF INTERNET AND SMART TECHNOLOGIES AND THEIR IMPACT ON SOCIETY

IoT is a type of technology that has an impact in almost all sectors, across countries across that are affected by COVID-19. IoT has taken over in almost all sectors rapidly, for instance, applications that are being used across the countries for tracking, tracing, and recording the data of the people using smartphones and having those applications in their smartphones and helping them become aware of their own conditions and those around them. Yet, if we dig deeper into the number of people using the benefits of IoT services, we might find it surprising that there are still places where there are either restrictions or there are some problems with using IoT; thus, some people are not able to make use of this technology. The biggest problem among many is network availability. There are places in India where there are network issues or only select applications can be used. There are people who are still unaware of the technologies that can benefit them, due to a lack of awareness and advertisement. There are still many people who cannot make payments online, which is much safer in the ongoing pandemic, and there are people who still do not have credit cards because they are not aware of the usage and privileges. In today's time, when classes are being provided online to follow social distancing, there are some rural areas where children are not able to take classes due to the unavailability of smartphones or due to network errors. Similarly, it can also be seen that there are some people who can work from home using smart technologies, yet there are some who are either deprived of smart technology or are unaware of it and so cannot use it for themselves. It is very important that continuous upgrades in the IoT sector spread awareness and make people aware of the privileges and also make them learn how to use such technologies for their own benefits.

18.2.4   Impact of COVID-19 on IoT Applications

18.2.4.1   Remote Asset Access

There are new and some existing IoT-based apps that have helped people stay in contact, conduct business meetings, and helped in taking online classes. IoT-based other software has helped in keeping in contact in a time of social distancing.

18.2.4.2   Digital Twin

In this time of COVID-19, companies are affected by production, shipping, and distribution delays as well as demand variability. This has led to the need for digital twins to create digital representations of an end-to-end supply chain that helps customers to explore dynamic sourcing options, evaluate risks, and evaluate tradeoffs to expedite decisions.

18.2.4.3   IoT Health Application Surge

COVID-19 has made it difficult for people to go out and visit doctors. In this time, telehealth consultations have helped people talk to doctors and receive advice. Digital diagnostics is another use of IoT e.g. digital thermometers saw a rise in use as COVID-19 spread in the U.S. [84]

18.2.4.4   Technologies Taking-Off

The use of technology has risen since the spread of COVID-19. The changes include working from home, collaboration tools, VPN networking, mobile devices, and security. Apps such as YeeCall, Bigo Live, Skype, WhatsApp, IMO, Facebook Messenger, FaceTime, Google Duo, and WebEx Meet are some that are used for video calling and meeting [85].

18.3 SMART TRACING TECHNOLOGIES

18.3.1 CONTACT TRACING TECHNOLOGIES (FIGURE 18.1) – REVIEW

Contact tracing is a way of tracking if a person has come in contact with someone who came in contact with an already infected person and for further collection of information about these contacts. The impact and effectiveness of technologies, such as smartphones and smartwatches with GPS, Bluetooth, cellular network, and Wi-Fi on the preciseness of the contract tracing on spreading and control of diseases are discussed. A model is built to analyze and check the capability and cost of measures to be taken, based on the smartphone contact tracing technology used. The result of the study was that to collect the data and trace COVID-19 infected persons, the technology needs to be precise; further, the contact needs to be traced quickly and a large number of people must be using the contact tracing application. Mobile phones that are used for contact tracing have a huge impact on social and economic value. Hence, contact tracing using technologies is effective [86–88]. The smartphone magnetometer, where two phones are present in a short distance showing a high linear correlation is discussed. It was studied how effective a diagnostic test is by evaluating it on the basis of differentiating and forecasting the capability of smartphone magnetometer-based contact detection [89]. On the basis of the tests, it was concluded that the smartphone meters were more efficient [90]. The different areas and aspects of development and the challenges of present contact tracing technology for individual and group contact tracing, static, and dynamic group contact tracing is studied. Multi-view contact tracing, multi-scale contact tracing, and AI-based contact tracing are provided for next-generation technologies for epidemic prevention and control. Advanced AI-based contact tracing methods will work efficiently in controlling the situation [91]. The student contacts within the disease propagation distance, then the construction of a disease propagation graph to model the infectious disease propagation is analyzed. On the basis of the graph, a metric called connectivity centrality to measure a node's importance was studied. The proposal discussed an efficient privacy-preserving contact tracing for infection detection by the help of which the users can safely upload their data to the server and later if any person is affected with the infection then the other users can check if they have ever been in contact with the person or not. A matching score is used for showing the results of tracing contact and other measures used. An adaptive scanning method to optimize the power consumption of the wireless scanning process has been developed [92].

A smart consumer electronics solution to provide and promote safe and step-by-step opening after the lockdown is finished. The easy band is introduced to restrict the growth of new cases with the help of auto contact tracing and by promoting the importance of social distancing [93]. The study includes how to reduce the tracing time and increase the accuracy of infection using Mobile Health Services (MHS). It is also studied that the correct location and bio-information of a person can be identified using RFID. The proposed model works for the tracing of the persons affected from different areas and records the data for future purposes. The steps taken by the government in Singapore, the abled agencies, and measures were taken

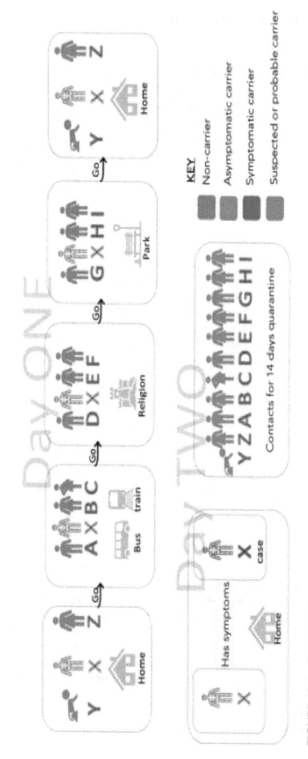

FIGURE 18.1 Contact tracing.

for the controlling and tracking of the emerging crisis [94]. The spread of COVID-19 and also the objective and need for contact tracing are studied. Severe Acute Respiratory Syndrome (SARS) cases are discussed, along with the detection of the disease [95,96]. The discrete event model of the response of the local public health department is modeled and further transferred between providers, laboratories, and LHDs, and examines the effect of different alerting strategies on the number of confirmed cases encountered. The effect of limited resource availability for contact tracing and resource availability has a significant impact on the progression of a disease outbreak, as do information delays at various stages of the process [97]. The community-based framework will help in tracking and recording the information about the contacts and on the basis of that, the message will be sent as an alert and hence will have to take vaccination [47]. The spread of false reports, mis-information, and unsolicited fear of COVID-19 have an impact on the world economy. The role, use, and importance of the technologies such as the IoT, un-manned aerial vehicles (UAVs), blockchain, AI, and 5G, among others, are dis-cussed, to minimize the effect of COVID-19 until any vaccine is found on a global level [98]. The virus spreading COVID-19 in all countries is discussed and there is a need for a smart and fast tracing system is proposed. A system is designed which can trace COVID by the thermal image with less human contact using IoT. [18,99,100].

18.3.2 POPULAR APPLICATIONS FOR CONTACT TRACING

It is known that, in India, more than 19 applications are being used to track COVID-19, among which the most popularly known is the Arogya Setu. The government in India has asked people to download the applications for the purpose of tracing the coronavirus-affected people in and around the location of the people who are downloading them, which will help the citizens track and be updated about the situation in their locality, in the nearby surroundings at different spatial distribution. The Arogya Setu app has been downloaded more than 10 million times from Android and IOS stores after it was launched on April 2, 2020. The location of the application is taken through the Global Positioning System (GPS), Bluetooth, and the contact number of the users. The information is afterward cross-referenced with the Indian Council of Medical Research (ICMR), in which the positive cases have been reported [101].In Kerala, the application used for tracking quarantined in-dividuals is Innefu's Unmaze app. The application has been used since March 25, 2020. Another application being used by the Kerala government is GoK Direct – Kerala, for providing recent health updates. The IT ministry has launched a new application, called Covid-19 feedback application, which is used for surveying and collecting information about users if they have undergone any treatment or have got any tests done at hospitals; the objective of the application is to help the government locate places where testing must be provided or increased, and the treatment process should be provided or increased. The survey of India has also launched an application called: SAHYOG, which will complement the Arogya Setu application to trace contacts, provide awareness to the public, and for the purpose of self-assessment. The Chhattisgarh government is using an application, called

CG Covid-19 ePass, to streamline the applications for the ePass for the purpose of the movement of vehicles between and within the district. The application has been downloaded over 50,000 times from Google Play. In Karnataka, the revenue department has made a Quarantine Watch, which is compulsory for people who are home quarantined. It used for the purpose of sending their details, which also includes a selfie every hour except between 22:00 hours and 07:00 hours. The application has been downloaded more than 100,000 times. Likewise, Surat Municipal Corporation, Gujrat developed the SMC COVID-19 Tracker with the objective of keeping a check on the people who are on home quarantined; the app also requires submission of selfies, and users are required to click on a button to provide their location every hour. The Maharashtra government launched Mahakavach for the purpose of tracing contact and keeps track of the people in quarantine. Corona Mukt Himachal is another application used by Himachal Pradesh to track home quarantined people. Tamil Nadu uses COVID-19 Quarantine Monitor Tamil Nadu application for tracking location.

18.3.3 OUTCOME OF CONTRACT TRACING APPLICATIONS

Contact tracing has been used for other diseases for tracing purposes, including tuberculosis, measles, HIV, bloodborne infections, and some dangerous bacterial infections and novel infections, such as SARS-CoV, H1N1, and COVID-19 [102]. The objectives of contact tracing are:

- To control the spread of infection from one person to another;
- To alter contacts to the possibility of infection and further provide required medication and care (Figure 18.2);
- To provide counseling and treatment after the test whether or not the test is positive;
- To prevent reinfection to the person who has already been infected once;
- To study the epidemiology of the disease in a particular population.

The complete eradication of smallpox was made possible with the help of contact tracing. Animals and individuals can wear a passive RFID tag without having mobile phones on them. To guarantee use, it is best used to access service while accumulating points. To the best of our knowledge, this is the first solution proposing IoT and specifically RFID for anonymized RFID contact tracing of infection spread (Figure 18.3).

18.3.4 POPULAR APPLICATIONS PAYMENT SYSTEM AND IMPACT

A contactless payment system refers to payment but without involving physical contact between the devices used in consumer payment and POS terminals by the merchant [103]. The most commonly used contactless methods are debit, credit, or smart cards, which can also be named as chip cards; these use RFID technology or NFC. The time required for the transaction to process differs from place to place, bank to bank. Some of the most commonly used NFC are Samsung Pay, Apple Pay,

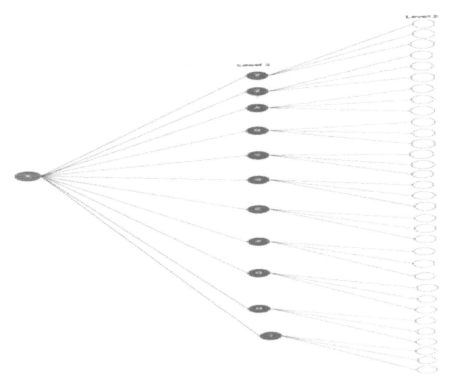

FIGURE 18.2 Exponential contacts and infection possibilities from index case.

Google Pay, Fitbit Pay, Phone Pe, BHIM, and all other applications which do not require any kind of contact. The USA, the UK, Japan, Germany, Canada, Australia, France, and the Netherlands, etc. are places where people make payments using credit cards, as it is an easy way of making payments and being contactless [104]. Over time, it can be seen that people have started using more online or cashless payment methods, as they are more convenient, but there are still many people, especially in rural areas, who are unaware of the benefits that they can have using the advanced technology systems helping in faster and safe way of making payments from anywhere to anywhere. The use of a cashless system can benefit in many ways like lesser crime rates, less time and cost, and easy currency exchange. But there are disadvantages too, sharing personal information, if hackers take away money, those who do not have bank accounts will not be able to comply with the changing cashless technology.

18.3.5 POPULAR APPLICATIONS FOR EDUCATION AND IMPACT

In today's time when the whole globe is affected by the COVID-19 pandemic, the system has evolved for the good of everyone. In this hard time, children cannot go to schools but education is important, so, to serve the purpose it can be seen that schools are using online modes for teaching students using different applications,

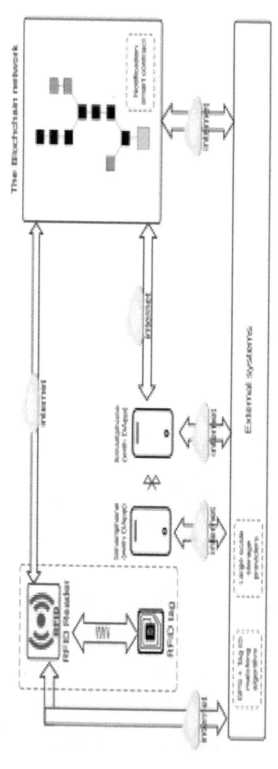

FIGURE 18.3 RFID device data flow diagram to the blockchain.

which can hold a number of students and teachers together for the class. There are websites that provide classes and tutorials for the students to understand the concepts. Some of the few websites are EDX: edx.org, Academic Earth, Internet archives, Big think, Coursera, Brightstrom, Cosmo learning, Futures Channel, and Howcast. Now, in India, there are many online teaching applications, including merit nation, byjus, my CBSE guide, vedantu, vidyakul, toppr, doubtnut, khan academy, unacademy, gradeup, etc. Despite the availability of so many online teaching platforms, there are parents and students who are still looking forward to school teaching and still are not able to adapt or comply with the online teaching system.

18.4   ANALYSIS AND SOCIAL IMPACT

The impact of technology in countries like India and across the globe has been seen for a long time but it has proved its worth majorly in the time of COVID-19. The different modified and advanced applications that are being used by people for different purposes have proved their worth but, at the same time, it is the need of the hour that people must also be aware of the usage of these applications and also how these applications will affect and benefit the people. The applications that are majorly being used are the self-surveillance applications with different purposes some help in getting information about the people who are infected by the virus in the nearby locality or area. Some applications are helping in contact tracing that is keeping a check on the medical history of the people one has met did ever suffer from any kind of disease and also from COVID-19 or not. Apart from the smart technologies of tracing and tracking corona-infected people and those with corona symptoms, there are also applications that have acted as a major support system in this difficult time when schools are not able to open in order to follow social distancing and save each other from getting affected by the virus. Schools are using different social platforms for keeping in touch with students and for conducting classes. There are applications that are providing a platform for people of different age groups to learn new things and do courses from their homes using smart technology. At the same time, it is important to mention that the COVID-19 virus spreads by the means of contact, and it has increased the need for using methods of payments that involve contactless payments, hence reducing the risk of spreading of the virus and also in helping consumers and producers to be in contact of each other and help the economy function. It is also important as one can make payments from one part of the country to any other part of the country in very little time. COVID-19 has proved the worth of the use of smart technologies and why it is important to invest more in the technology and bring new technology with advanced features and increased accuracy. It is the social media platform that has also played an important role in the awareness about the spreading of viruses, which could be only possible due to smart technologies and smartphones. At the same time, there are still some parts of the country where people are still not aware or able to use such technology or take the benefits of the available smart applications for their own benefit. The major reasons among many are network hindrance and the economic conditions of the majority of people in the rural areas. At the same time, there are disadvantages of mistrust expanding onto surveillance apps like Aarogya Setu and mistrust of technology to educate children in schools or contactless transactions (many cases of fraudulent transactions registered during COVID-19 through emails from imposters providing free COVID tests). On the other hand, ed-tech companies and apps are accommodating the shifts and gaps between students and tech through platforms like massive open online courses (MOOCs) and remote educational sources. Primary education is yet to have a grasp of education over the Internet. Primary educators are not well-equipped and familiar with ed-tech platforms, hence impacting their delivery. Students are in shock with a shift from pencil-paper method to Zoom, WhatsApp evaluations, also making it less feasible for students in remote areas to access these platforms altogether. Music has become the universal language of the pandemic in COVID. People singing in their balconies to document provided by songs on COVID virus. Internet research furthering the global transfer of knowledge and products making the technology solutions more people-oriented than economy-oriented. Despite the fact that contact tracing is the most used method to keep a track of the spreading of disease, at the same time, contact tracing is an important tool for investigating any new disease or any unexpected outbreaks. Examples are as follows, in the case of SARS, contact tracing has been used for the purpose of recording if probable cases are connected to some preexisting cases of disease and further to record if any secondary transmission has taken place or not in a specific community. Contact tracing was also implemented at the time of H1N1 influenza in the year 2009 among flight passengers [105]. Although it was very difficult to achieve the goal in such a chaotic situation, there has always been renovation in the guidelines and strategies for improving efficiency. In the year 2020 on April 10, Apple and Google, which are the most used brands of mobile phones, announced a new technology for the purpose of tracking COVID-19 in both iOS and Android. Relying on Bluetooth low-energy wireless radio signals for the objective of tracing contact, these new technologies would help the users by providing them information about the people if they have been in contact with, who would be suffering from the COVID-19 virus.

18.4.1 Privacy Concern

Different rules and regulations, which are Pan-European privacy-preserving proximity tracing (PEPP-PT), whisper tracing protocol, decentralized privacy-preserving proximity tracing (DP-PPT/DP-3T), TCN protocol, contact event numbers (CEN), privacy sensitive protocols, and mechanisms for mobile contact tracing (PACT), and others, were mentioned to maintain the privacy of the users [99].

18.4.2 User Adoption, Etc.

The major issue with the contact tracing application is about medical privacy and confidentiality. To achieve the object of exact contact tracing and tracking the person who might be in contact with the person who was already affected by the

COVID-19, medical practitioners cannot maintain the confidentiality of users. Also, people who have been infected by the virus are not told about the person from whom they got the infection or virus. As a result of which, most people are hesitant to seek medical help as their privacy is exposed, so they fear discrimination and abuse. The contact tracing health officials need to understand the importance of trust along with vulnerable populations and sensitivity toward an individual's condition or situation.

18.5 CONCLUSION

IoT-based applications have shown considerable effectiveness in different sectors, especially in the healthcare sector. For the time that countries have been affected by COVID-19, the use of IoT has increased. Many countries have set up temperature measurement systems at the entrance of public transport. By collecting all this data through a network, a real-time study can be conducted, which will help in the better study of COVID-19.

REFERENCES

1. "Contact Tracing – Bluetooth Specification" (PDF). covid19-static.cdn-apple.com (Preliminary ed.). 2020-04-10. Archived (PDF) from the original on 2020-04-10. Retrieved 2020-04-10.
2. "Severe Acute Respiratory Syndrome (SARS) – multi-country outbreak". Global Alert and Response. WHO. Retrieved 2013-05-28.
3. A. K. Tripathy, A. G. Mohapatra, S. P. Mohanty, E. Kougianos, A. M. Joshi and G. Das, "EasyBand: A wearable for safety-aware mobility during pandemic outbreak," *IEEE Consumer Electronics Magazine*, doi: 10.1109/MCE.2020.2992034
4. https://en.wikipedia.org/wiki/List_of_epidemics
5. https://covid19.who.int/
6. Peter Mitchell, "The archaeological study of epidemic and infectious disease," *World Archaeology*, vol. 35, no. 2, pp. 171–179, 2003, DOI: 10.1080/0043824032 000111353
7. https://www.webmd.com/cold-and-flu/what-are-epidemics-pandemics-outbreaks
8. G. Stewart, K. Heusden and G. A. Dumont, "How control theory can help us control Covid-19," *IEEE Spectrum*, vol. 57, no. 6, pp. 22–29, June 2020, doi: 10.1109/ MSPEC.2020.9099929.
9. I. Monteiro, E. Rocha, E. Silva, G. L. Santos, W. Santos and P. T. Endo, "Developing an e-Health system based on IoT, fog and cloud computing," 2018 IEEE/ACM International Conference on Utility and Cloud Computing Companion (UCC Companion), Zurich, 2018, pp. 17–18, doi: 10.1109/UCC-Companion.201 8.00024.
10. J. Jiang et al., "A distributed RSS-based localization using a dynamic circle expanding mechanism," *IEEE Sensors Journal*, vol. 13, no. 10, pp. 3754–3766, Oct. 2013, doi: 10.1109/JSEN.2013.2258905.
11. M. Yannuzzi, R. Milito, R. Serral-Gracià, D. Montero and M. Nemirovsky, "Key ingredients in an IoT recipe: Fog computing, cloud computing, and more fog computing," 2014 IEEE 19th International Workshop on Computer Aided Modeling and Design of Communication Links and Networks (CAMAD), Athens, 2014, pp. 325–329, doi: 10.1109/CAMAD.2014.7033259.

12. S. Ramnath, A. Javali, B. Narang, P. Mishra and S. K. Routray, "IoT based localization and tracking," 2017 International Conference on IoT and Application (ICIOT), Nagapattinam, 2017, pp. 1–4, doi: 10.1109/ICIOTA.2017.8073629.

13. W. Xi and L. Ling, "Research on IoT privacy security risks," 2016 International Conference on Industrial Informatics – Computing Technology, Intelligent Technology, Industrial Information Integration (ICIICII), Wuhan, 2016, pp. 259–262, doi: 10.1109/ICIICII.2016.0069.

14. Chen, Z., F. Xia, T. Huang et al., "A localization method for the Internet of Things," *Journal of Supercomputing*, vol. 63, no. 657–674 (2013). https://doi.org/10.1007/s11227-011-0693-2.

15. Raju Vaishyaa, Mohd Javaidb, Ibrahim Haleem Khan, "Artificial Intelligence (AI) applications for COVID-19 pandemic," *Diabetes & Metabolic Syndrome: Clinical Research & Reviews*, vol. 14, no. 4, pp. 337–339, 2020.

16. Ravi Pratap Singh, Mohd Javaid, Abid Haleem and Rajiv Suman, "Internet of things (IoT) applications to fight against COVID-19 pandemic," *Diabetes & Metabolic Syndrome: Clinical Research & Reviews*, vol. 14, no. 4, pp. 521–524, July–August 2020.

17. Lu Tan and Neng Wang, "Future internet: The Internet of Things," 2010 3rd International Conference on Advanced Computer Theory and Engineering(ICACTE), Chengdu, 2010, pp. V5–376-V5–380, doi: 10.1109/ICACTE.2010.5579543.

18. A. Zanella, N. Bui, A. Castellani, L. Vangelista and M. Zorzi, "Internet of things for smart cities," *IEEE Internet of Things Journal*, vol. 1, no. 1, pp. 22–32, Feb. 2014, doi: 10.1109/JIOT.2014.2306328.

19. Abdulrazaq, Assoc. Prof. Dr. Mohammed, Nurul Hazairin, Zuhriyah Halimatuz, Salah Al-Zubaidi, Sairah Karim, Safinaz Mustapha and Eddy Yusuf. (2020). 2019 Novel Coronavirus Disease (Covid-19): Detection and Diagnosis System Using IoT Based Smart Glasses. 29.

20. C. Perera, A. Zaslavsky, P. Christen and D. Georgakopoulos, "Context Aware Computing for The Internet of Things: A Survey," in *IEEE Communications Surveys & Tutorials*, vol. 16, no. 1, pp. 414–454, First Quarter 2014, doi: 10.1109/SURV.2013.042313.00197.

21. E. Welbourne et al., "Building the Internet of Things Using RFID: The RFID Ecosystem Experience,"in *IEEE Internet Computing*, vol. 13, no. 3, pp. 48–55, May–June 2009, doi: 10.1109/MIC.2009.52.

22. R. Nukala, K. Panduru, A. Shields, D. Riordan, P. Doody and J. Walsh, "Internet of Things: A review from 'Farm to Fork,'" 2016 27th Irish Signals and Systems Conference (ISSC), Londonderry, 2016, pp. 1–6, doi: 10.1109/ISSC.2016.7528456.

23. Al-Fuqaha, M. Guizani, M. Mohammadi, M. Aledhari and M. Ayyash, "Internet of things: A survey on enabling technologies, protocols, and applications," *IEEE Communications Surveys & Tutorials*, vol. 17, no. 4, pp. 2347–2376, 2015, doi: 10.1109/COMST.2015.2444095.

24. R. Khan, S. U. Khan, R. Zaheer and S. Khan, "Future internet: The internet of things architecture, possible applications and key challenges," 2012 10th International Conference on Frontiers of Information Technology, Islamabad, 2012, pp. 257–260, doi: 10.1109/FIT.2012.53.

25. Y. Huang and G. Li, "A semantic analysis for internet of things," 2010 International Conference on Intelligent Computation Technology and Automation, Changsha, 2010, pp. 336–339, doi: 10.1109/ICICTA.2010.73.

26. B. Nath, F. Reynolds and R. Want, "RFID technology and applications," *IEEE Pervasive Computing*, vol. 5, no. 1, pp. 22–24, Jan–March 2006. doi: 10.1109/MPRV.2006.13.

27. P. Kumari, M. López-Benítez, G. M. Lee, T. Kim and A. S. Minhas, "Wearable internet of things - From human activity tracking to clinical integration," 2017 39th Annual

International Conference of the IEEE Engineering in Medicine and Biology Society (EMBC), Seogwipo, pp. 2361–2364, 2017. doi: 10.1109/EMBC.2017.8037330.

28. M. Desai and A. Phadke, "Internet of things based vehicle monitoring system," 2017 Fourteenth International Conference on Wireless and Optical Communications Networks (WOCN), Mumbai, 2017, pp. 1–3, doi: 10.1109/WOCN.2017.8065840.

29. I. Stojmenovic, "Fog computing: A cloud to the ground support for smart things and machine-to-machine networks," 2014 Australasian Telecommunication Networks and Applications Conference (ATNAC), Southbank, VIC, pp. 117–122, 2014. doi: 10.1109/ATNAC.2014.7020884.

30. Md Siddikur Rahman, Noah C Peeri, Nishtha Shrestha, Rafdzah Zaki, Ubydul Haque and Siti Hafizah Ab Hamid, "Defending against the Novel Coronavirus (COVID-19) outbreak: How can the internet of things (IoT) help to save the world?" *Health Policy and Technology, 2020.* 10.1016/j.hlpt.2020.04.005.

31. Abid Haleem et al., "Effects of COVID 19 pandemic in daily life," *Current Medicine Research and Practice,* vol. 10, no. 2, pp. 78–79, 3 Apr. 2020. doi: 10.101 6/j.cmrp.2020.03.011

32. S. Andreev, C. Dobre and P. Misra, "Internet of things and sensor networks," *IEEE Communications Magazine,* vol. 58, no. 4, pp. 74–74, April 2020. doi: 10.1109/ MCOM.2020.9071994.

33. V. Chamola, V. Hassija, V. Gupta and M. Guizani, "A comprehensive review of the COVID-19 pandemic and the role of IoT, drones, AI, blockchain, and 5G in managing its impact," *IEEE Access,* vol. 8, pp. 90225–90265, 2020. doi: 10.1109/ ACCESS.2020.2992341

34. X. Pan, Application of personal-oriented digital technology in preventing transmission of COVID-19, China. *Irish Journal of Medical Science, 2020.* https:// doi.org/10.1007/s11845-020-02215-5.

35. Sandeep K. Sood and Isha Mahajan, "Computers in industry: *Wearable IoT sensor-based healthcare system for identifying and controlling chikungunya virus,"* volume 91, October 2017, pp. 33–44.

36. F. Shi et al., "Review of artificial intelligence techniques in imaging data acquisition, segmentation and diagnosis for COVID-19," *IEEE Reviews in Biomedical Engineering,* doi: 10.1109/RBME.2020.2987975.

37. Deblina Roy, Sarvodaya Tripathy, Sujita Kumar Kar, Nivedita Sharma, Sudhir Kumar Verma and Vikas Kaushal, "Study of knowledge, attitude, anxiety & perceived mental healthcare need in Indian population during COVID-19 pandemic," *Asian Journal of Psychiatry,* vol. 51, June 2020. DOI: 10.1016/j.ajp.2020.102083

38. A. Srinivasa Rao and J. Vazquez, "Identification of COVID-19 can be quicker through artificial intelligence framework using a mobile phone-based survey when cities and towns are under quarantine," *Infection Control & Hospital Epidemiology,* vol. 41, no. 7, pp. 826–830, 2020. doi: 10.1017/ice.2020.61,

39. R. F. Sear et al., "Quantifying COVID-19 content in the online health opinion war using machine learning," *IEEE Access,* vol. 8, pp. 91886–91893, 2020, doi: 10.11 09/ACCESS.2020.2993967.

40. L. Li et al., "Characterizing the propagation of situational information in social media during COVID-19 epidemic: A case study on Weibo," *IEEE Transactions on Computational Social Systems,* vol. 7, no. 2, pp. 556–562, April 2020, doi: 10.1109/ TCSS.2020.2980007.

41. A. A. R. Alsaeedy and E. K. P. Chong, "Detecting regions at risk for spreading COVID-19 using existing cellular wireless network functionalities," in *IEEE Open Journal of Engineering in Medicine and Biology,* vol. 1, pp. 187–189, 2020, doi: 10.1109/OJEMB.2020.3002447.

42. F. Rustam et al., "COVID-19 future forecasting using supervised machine learning models," *IEEE Access*, vol. 8, pp. 101489–101499, 2020, doi: 10.1109/ACCESS.2 020.2997311.

43. M. A. Mahmud, K. Bates, T. Wood, A. Abdelgawad and K. Yelamarthi, "A complete internet of things (IoT) platform for structural health monitoring (SHM)," 2018 IEEE 4th World Forum on Internet of Things (WF-IoT), Singapore, 2018, pp. 275–279, doi: 10.1109/WF-IoT.2018.8355094.

44. N. Zheng et al., "Predicting COVID-19 in China using hybrid AI model," *IEEE Transactions on Cybernetics*, vol. 50, no. 7, pp. 2891–2904, July 2020. doi: 10.11 09/TCYB.2020.2990162.

45. D. Dong et al., "The role of imaging in the detection and management of COVID-19: A review," *IEEE Reviews in Biomedical Engineering*. doi: 10.1109/RBME.202 0.2990959.

46. M. Quayson, C. Bai and V. Osei, "Digital inclusion for resilient post-COVID-19 supply chains: Smallholder farmer perspectives," in *IEEE Engineering Management Review*. doi: 10.1109/EMR.2020.3006259.

47. Y. Ren, J. Yang, M. C. Chuah and Y. Chen, "Mobile phone enabled social community extraction for controlling of disease propagation in healthcare," 2011 IEEE Eighth International Conference on Mobile Ad-Hoc and Sensor Systems, Valencia, 2011, pp. 646–651. doi: 10.1109/MASS.2011.68

48. T. Worth, R. Uzsoy, E. Samoff, A. Meyer, J. Maillard and A. M. Wendelboe, "Modelling the response of a public health department to infectious disease," Proceedings of the 2010 Winter Simulation Conference, Baltimore, MD, 2010, pp. 2185–2198. doi: 10.1109/WSC.2010.5678917

49. "IEEE Smart Grid Vision for Vehicular Technology: 2030 and Beyond Roadmap," in *IEEE Smart Grid Vision for Vehicular Technology: 2030 and Beyond Roadmap*, pp. 1–17, 30 June 2015, doi: 10.1109/IEEESTD.2014.6716939.

50. F. M. Bhutta, "Application of smart energy technologies in building sector – future prospects," 2017 International Conference on Energy Conservation and Efficiency (ICECE), Lahore, pp. 7–10, 2017. doi: 10.1109/ECE.2017.8248820.

51. F. Wu, T. Wu and M. R. Yuce, "Design and implementation of a wearable sensor network system for IoT-connected safety and health applications," 2019 IEEE 5th World Forum on Internet of Things (WF-IoT), Limerick, Ireland, pp. 87–90, 2019. doi: 10.1109/WF-IoT.2019.8767280.

52. Y. Mano et al., "Exploiting IoT technologies for enhancing Health Smart Homes through patient identification and emotion recognition," *Computer Communications*, vol. 89, no. 90, pp. 178–190, 2015.

53. G. Yang et al., "A Health-IoT platform based on the integration of intelligent packaging, unobtrusive bio-sensor, and intelligent medicine box," in *IEEE Transactions on Industrial Informatics*, vol. 10, no. 4, pp. 2180–2191, Nov. 2014, doi: 10.1109/TII.2014.2307795.

54. J. Cooper and A. James, "Challenges for database management in the internet of things," *IETE Technical Review*, vol. 26, no. 5, pp. 320–329, 2009.

55. S. S. Mishra and A. Rasool, "IoT health care monitoring and tracking: A survey," 2019 3rd International Conference on Trends in Electronics and Informatics (ICOEI), Tirunelveli, India, 2019, pp. 1052–1057, doi: 10.1109/ICOEI.2019.8862763.

56. V. Pardeshi, S. Sagar, S. Murmurwar and P. Hage, "Health monitoring systems using IoT and Raspberry Pi – A review," 2017 International Conference on Innovative Mechanisms for Industry Applications (ICIMIA), Bangalore, vol. 14, no. 4, 2020, pp. 134–137, 2017. doi: 10.1109/ICIMIA.2017.7975587.

57. C. Raj, C. Jain and W. Arif, "HEMAN: Health monitoring and nous: An IoT based e-health care system for remote telemedicine," 2017 International Conference on

Wireless Communications, Signal Processing and Networking (WiSPNET), Chennai, 2017, pp. 2115–2119, doi: 10.1109/WiSPNET.2017.8300134.

58. X. Sun, Z. Lu, X. Zhang, M. Salathé and G. Cao, "Targeted vaccination based on a wireless sensor system," 2015 IEEE International Conference on Pervasive Computing and Communications (PerCom), St. Louis, MO, pp. 215–220, 2015. doi: 10.1109/PERCOM.2015.7146531

59. A. A. Hussain, O. Bouachir, F. Al-Turjman and M. Aloqaily, "AI techniques for COVID-19," in *IEEE Access*, vol. 8, pp. 128776–128795, 2020, doi: 10.1109/ACCESS.2020.3007939.

60. G. B. Rehm et al., "Leveraging IoTs and machine learning for patient diagnosis and ventilation management in the intensive care unit," in *IEEE Pervasive Computing*, doi: 10.1109/MPRV.2020.2986767.

61. G. Laštovička-Medin, "Social engineering and prototype awareness enhancing during international coronavirus outbreak," 2020 9th Mediterranean Conference on Embedded Computing (MECO), Budva, Montenegro, pp. 1–4, 2020. doi: 10.1109/MECO49872.2020.9134355.

62. H. Lyu, L. Chen, Y. Wang and J. Luo, "Sense and sensibility: Characterizing social media users regarding the use of controversial terms for COVID-19," in *IEEE Transactions on Big Data*, doi: 10.1109/TBDATA.2020.2996401.

63. M. Abdel-Basset, R. Mohamed, M. Elhoseny, R. K. Chakrabortty and M. Ryan, "A hybrid COVID-19 detection model using an improved marine predators algorithm and a ranking-based diversity reduction strategy," in *IEEE Access*, vol. 8, pp. 79521–79540, 2020. doi: 10.1109/ACCESS.2020.2990893.

64. M. E. H. Chowdhury et al., "Can AI help in screening Viral and COVID-19 pneumonia?," in *IEEE Access*, doi: 10.1109/ACCESS.2020.3010287.

65. Singh, Ravi Pratap et al., "Internet of things (IoT) applications to fight against COVID-19 pandemic," *Diabetes & Metabolic Syndrome,* vol. 14, no. 4, pp. 521–524, 2020. doi: 10.1016/j.dsx.2020.04.041

66. A. Mazayev, J. A. Martins, and N. Correia, "Interoperability in IoT Through the Semantic Profiling of Objects," *IEEE Access*, vol. 6, pp. 19379–19385, 2018.

67. A. S. A. Daia, R. A. Ramadan, and M. B. Fayek, "Sensor networks attacks classifications and mitigation", *Annals of Emerging Technologies in Computing (AETiC)*, vol. 2, no. 4, pp. 28–43, 2018.

68. B. Buntz, "IoT data: Strategic sharing will yield big innovations," 20 6 2017. [Online]. Available: http://www.ioti.com/strategy/iot-data-strategic-sharing-will-yieldbig-innovations. [Accessed 22 7 2018].

69. D. B. Ansari, A.-U. Rehman, and R. Ali, "Internet of things (IoT) protocols: A brief exploration of MQTT and CoAP," *International Journal of Computer Applications*, vol. 179, pp. 9–14, 03 2018.

70. E. Borgia, D. G. Gomes, B. Lagesse, R. Lea, and D. Puccinelli, "Special issue on 'Internet of things: Research challenges and Solutions'," *Computer Communications*, vol. 89, no. 90, pp. 1–4, 2016.

71. G. Papadodimas, G. Palaiokrasas, A. Litke and T. Varvarigou, "Implementation of smart contracts for blockchain based IoT applications," 2018 9th International Conference on the Network of the Future (NOF), Poznan, 2018, pp. 60–67.

72. H. U. Rehman, M. Asif, and M. Ahmad, "Future applications and research challenges of IOT," in 2017 International Conference on Information and Communication Technologies (ICICT), pp. 68–74, Dec 2017. (IJACSA) International Journal of Advanced Computer Science and Applications, Vol. 10, No. 6, 2019

73. K. K. Patel, S. M. Patel, et al., "Internet of things IOT: definition, characteristics, architecture, enabling technologies, application future challenges," *International Journal of Engineering Science and Computing*, vol. 6, no. 5, pp. 6122–6131, 2016.

74. M. H. Miraz, M. Ali, P. S. Excell, and R. Picking, "A review on internet of things (IoT), internet of everything (IoE) and internet of nano things (IoNT)," 2015 Internet Technologies and Applications (ITA), pp. 219– 224, Sep. 2015.

75. M. Miraz, M. Ali, P. Excell, and R. Picking, "Internet of nano-things, things and everything: Future growth trends," *Future Internet*, vol. 10, no. 8, p. 68, 2018.

76. P. J. Ryan and R. B. Watson, "Research challenges for the internet of things: What role can or play?," *Systems*, vol. 5, no. 1, pp. 1–34, 2017.

77. P. Tadejko, "Application of internet of things in logistics-current challenges," *EkonomiaiZarz{a,}dzanie*, vol. 7, no. 4, pp. 54–64, 2015.

78. Q. Pham, D. C. Nguyen, T. Huynh-The, W. Hwang and P. N. Pathirana, "Artificial intelligence (ai) and big data for coronavirus (COVID-19) pandemic: A survey on the state-of-the-arts," in *IEEE Access*, vol. 8, pp. 130820–130839, 2020, doi: 10.11 09/ACCESS.2020.3009328.

79. R. Porkodi and V. Bhuvaneswari, "The internet of things (iot) applications and communication enabling technology standards: an overview," in 2014 International Conference on Intelligent Computing Applications, pp. 324–329, March 2014.

80. A.O. Salau, N. Marriwala, M. Athaee, Data Security in Wireless Sensor Networks: Attacks and Countermeasures. *Lecture Notes in Networks and Systems*, vol. 140. Springer, Singapore, 2021. DOI: 10.1007/978-981-15-7130-5_13

81. S. Rajguru, S. Kinhekar, and S. Pati, "Analysis of internet of things in a smart environment," *International Journal of Enhanced Research in Man-agreement and Computer Applications*, vol. 4, no. 4, pp. 40–43, 2015.

82. S. V. Zanjal and G. R. Talmale, "Medicine reminder and monitoring system for secure health using IOT," *Procedia Computer Science*, vol. 78, pp. 471–476, 2016.

83. E. P. Yadav, E. A. Mittal and H. Yadav, "IoT: Challenges and issues in Indian perspective," in 2018 3rd International Conference On Internet of Things: Smart Innovation and Usages (IoT-SIU), Bhimtal, 2018, pp. 1–5, doi: 10.1109/IoT-SIU.2018.8519869.

84. Cohen, I. G., Gostin, L. O., and Weitzner, D. J. "Digital smartphone tracking for COVID-19: public health and civil liberties in tension," *JAMA*, vol. 323, no 23, pp. 2371–2372, 2020.

85. Bradford, L., Aboy, M. and Liddell, K., "COVID-19 contact tracing apps: a stress test for privacy, the GDPR, and data protection regimes," Journal of Law and the Biosciences, 7(1), p.lsaa034, 2020.

86. Anita Ramsetty, and Cristin Adams, "Impact of the digital divide in the age of COVID-19," *Journal of the American Medical Informatics Association*, ocaa078, https://doi.org/10.1093/jamia/ocaa078

87. E. Hernández-Orallo, P. Manzoni, C. T. Calafate and J. Cano, "Evaluating how smartphone contact tracing technology can reduce the spread of infectious diseases: The case of COVID-19," in *IEEE Access*, vol. 8, pp. 99083–99097, 2020, doi: 10.11 09/ACCESS.2020.2998042

88. F. Noonan, S. Maharjan, A. Shields, D. Riordan, J. Walsh and P. Doody, "Automation of a paper-based waste tracking system," 2018 2nd International Symposium on Small-Scale Intelligent Manufacturing Systems (SIMS), Cavan, 2018, pp. 1–5, doi: 10.1109/SIMS.2018.8355290.

89. Justin Chan et al. "PACT: Privacy sensitive protocols and mechanisms for mobile contact tracing". arXiv:2004.03544v1 [cs.CR], April 7, 2020.

90. S. Jeong, S. Kuk and H. Kim, "A smartphone magnetometer-based diagnostic test for automatic contact tracing in infectious disease epidemics," in *IEEE Access*, vol. 7, pp. 20734–20747, 2019. doi: 10.1109/ACCESS.2019.2895075

91. H. Chen, B. Yang, H. Pei and J. Liu, "Next generation technology for epidemic prevention and control: Data-driven contact tracking," in *IEEE Access*, vol. 7, pp. 2633–2642, 2019. doi: 10.1109/ACCESS.2018.2882915

92. Cheng-Ju Li et al., Mobile healthcare service system using RFID. In Conference Proceeding – 2004 IEEE International Conference on Networking, Sensing and Control, vol. 2, pp. 1014–1019, 2004.

93. Altuwaiyan Thamer, Hadian Mohammad and Liang Xiaohui, "EPIC: Efficient Privacy-Preserving Contact Tracing for Infection Detection," 1–6, 2018. 10.1109/ICC.2018.8422886.

94. P. R. Devadoss, Shan Ling Pan and S. Singh, "Managing knowledge integration in a national health-care crisis: Lessons learned from combating SARS in Singapore," in *IEEE Transactions on Information Technology in Biomedicine*, vol. 9, no. 2, pp. 266–275, June 2005. doi: 10.1109/TITB.2005.847160

95. Luca Ferretti et al., "Quantifying SARS-CoV-2 transmission suggests epidemic control with digital contact tracing," *Science,* vol. 368, no. 6491, pp. eabb6936, March 31, 2020. doi:10.1126/science.abb6936.

96. K. Leong, Y. Si, R. P. Biuk-Aghai and S. Fong, "Contact tracing in healthcare digital ecosystems for infectious disease control and quarantine management," 2009 3rd IEEE International Conference on Digital Ecosystems and Technologies, Istanbul, 2009, pp. 306–311, doi: 10.1109/DEST.2009.5276730

97. D. S. W. Ting, L. Carin, V. Dzau et al., "Digital technology and COVID-19," *Nat Med, vol.* 26, pp. 459–461, 2020. https://doi.org/10.1038/s41591-020-0824-5.

98. T. Tamai, Y. Saitoh, S. Sawada and Y. Hattori, "Peculiarities characteristics between contact trace and contact resistance of tin plated contacts," 2008 Proceedings of the 54th IEEE Holm Conference on Electrical Contacts, Orlando, FL, 2008, pp. 337–343. doi: 10.1109/HOLM.2008.ECP.65.

99. Natasha Lomas, "An EU coalition of techies is backing a 'privacy-preserving' standard for COVID-19 contacts tracing," TechCrunch, 2020-04-01. Archived from the original on 2020-04-10. Retrieved 2020-04-11.

100. Natasha Lomas, "EU privacy experts push a decentralized approach to COVID-19 contacts tracing," TechCrunch, 2020-04-06. Archived from the original on 2020-04-10. Retrieved 2020-04-11.

101. https://theprint.in/india/india-has-at-least-62-apps-to-deal-with-covid-but-they-all-do-the-same-job-mostly/439040/

102. Ontario Provincial Infectious Diseases Advisory Committee, *Sexually transmitted infections best practices and contact tracing best practice recommendations.* Toronto, Canada: Ontario Ministry of Health and Long-Term Care, 2009. ISBN 978-1-42497946-2.

103. Ian Sherr and Richard Nieva, "Apple and Google are building coronavirus tracking tech into iOS and Android – The two companies are working together, representing most of the phones used around the world". CNET, 2020-04-10. Archived from the original on 2020-04-10. Retrieved 2020-04-10.

104. Z. Han et al., "Accurate screening of COVID-19 using attention based deep 3D multiple instance learning," *IEEE Transactions on Medical Imaging*, doi: 10.1109/TMI.2020.2996256.

105. Corien M Swaan, Rolf Appels, Mirjam EE Kretzschmar, and Jim E van Steenbergen, "Timeliness of contact tracing among flight passengers for influenza A/H1N1 2009," *BMC Infectious Diseases* , vol.11, no. 1, pp. 355, 2011. doi:10.1186/1471-2334-11-355. ISSN 1471-2334. PMC 3265549. PMID 22204494.

19 Agriculture-Internet of Things (A-IoT)

Key Roles in Addressing Some Challenges in Agriculture

Ibrahim Muhammad Abdul
Nigerian Stored Products Research Institute, Kano

19.1 INTRODUCTION

The term "Internet of Things" was used during the launching of Radio-Frequency Identification (RFID) by Kevin Ashton in 1999 for the supply chain management. Breakthrough IoT technologies have diversified and can link different objects within the frame of a network without human intervention to commit error [1]. More simply, a physical object connected to the Internet through the use of an embedded device to monitor and control applications can be an IoT device. The concept of the IoT in an agricultural context refers to the connections of physical objects (such as animals, tractors, wire fences, storage, etc.) and the Internet through wireless networks with minimum or no human interventions. In recent years, IoT has found a wider application in the area of agriculture where it can deliver quality services at a lower cost. As a result of its wide abilities in sensing, tracking, connecting, identification, and analyzing data from objects, the exponential growth of the agriculture IoT market is projected to grow from USD 12.7 billion through 2019 to USD 20.9 billion by 2024 [2].

As a field of concern, agriculture is one of the fundamental sectors expected to be notably stimulated with the use of IoT. Looking at the current statistics of the world population (7.6 billion) as of 2017, the department of economic and social affairs of the United Nation (UN) projected that there will be exponential growth of world population, the world human populace is estimated to reach 8.6 billion in 2030, and 9.8 billion in 2050. Around 28.95% is expected to be added to the world population by 2050. To provide and feed such a target population globally, there is a need to increase the output of food production by 70%, and this must be achieved by 2050. As a result of the supply shortages of labor force and the high demands for food security, the adoption of IoT technology by farmers and growers for modernization and intensification of agricultural practices has been a target. At the same time, the

DOI: 10.1201/9781003140443-19

rising call for high-quality products together with the need for efficient use of water, nutrient content, and other natural resources call for a great concern [3].

Farmers and agricultural companies in developed and developing countries, such as the USA, Germany, Malaysia, China, and India, etc. are adopting IoT-based technologies for enhancing their production and gaining larger market shares. This chapter focused on looking at the basic concept, architecture, and the potential roles that the IoT plays in the agricultural sector [4,5].

19.2 BACKGROUND AND CONCEPT OF IOT

Generally, the IoT can be described as the inter-connection of physical objects with the resultant exchange of data over the cloud by the use of embedded technologies, such as sensors, routers, actuators, camera devices, software, and other sophisticated tools. The three simplistic layer structures deploy for IoT consist of the Sensing layer, the Network layer, and the Application layer as shown in Figure 19.1. The IoT is one of the fourth (4th) industrial revolution technologies that has hit the ground. It is efficient, and most importantly, a technique that was developed to solve problems. The IoT evolved from exclusive building blocks that incorporate masses of sensors, software programs, network components, and different digital gadgets. Also, the IoT permits the exchange of data via smart getaways with no human interventions. The efficiency of IoT technology generally lies on reasons of:

1. Intelligent and sensing abilities
2. Minimal human intervention
3. Faster access to data
4. Good time and cost management
5. Efficient communication

Nowadays, many industries in the agricultural sector have turned toward adopting IoT technology for smart farming, food grains storage, and food processing to enhance productivity and efficiency at production, processing, as well as the marketing of produce. Notwithstanding, advanced IoT enables work to be done faster and more accurately without error committed by man. The technological

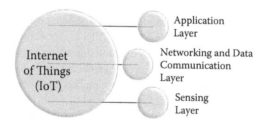

FIGURE 19.1 Process chart of three-layer IoT architecture in agriculture (Source: Own chart).

advancement in the field ensures the development of sensors and processors that are easy to carry, more sophisticated, socially acceptable, and economically viable. The communication networks are easily accessed worldwide so that smart farming can be achieved with a full pledge. Looking at the current agricultural practices in many developing countries, such as the use of cutlass, hoe, diggers, planter, harrower, and other simple tools, planting time, fertilizer requirement, and herbicide application, encouraging technologies such as smart farming or modern agriculture with the use of sophisticated equipment or IoT devices is the way forward to the problems malingering the industry.

19.3   IOT ARCHITECTURE IN AGRICULTURE

The IoT architecture known as enabling technologies may be sub-classed into three (3), the sensing layer, the network and data communication layer, and the application layer. It goes beyond Machine-to-Machine (M2M) scenarios and gives connectivity access between users, and without the presence of these three layers, IoT cannot be considered as a complete structure. The layers along with their functions are explained as follows;

The **sensing layer,** also known as the device layer, is one of the functional component or part of the IoT structure that enables for (i) automatic identification (ii) sensing, and (iii) bodily gadgets actuating (e.g. plant, parcel, animal, stable, pallet, truck, container) or acquisition of raw data from the surroundings. The agricultural sector made use of vital automatic identification (AutoID) technologies including barcodes and Radio-Frequency Identification (RFID). RFID technology has wide applications in the field of livestock management for animals' identification and supply chain management to monitor farm products during transportation. In open-field agriculture, sensors are needed in collecting data from the surrounding area. The data collected are recorded and further analyzed by actuators. Actuators measure to provides real-time feedback to the agronomist or farmer.

Network and data communication layer: The IoT devices are connected to the cloud through a very complex energy demand communication stack, called Internet Protocol. It also utilizes the non-IP networks via Internet gateway in communicating and transporting seamless data generated from the sensing layer, using intelligent connectivity technologies such as Bluetooth, RFID, Wi-Fi, and NFC. The leading network and data communication technologies like Sigfox, ISA100.11a, 6LowPan, ONE-NET, WirelessHART, Zigbee, are widely used in IoT. Wireless Sensor Networks (WSNs) in recent years, is applied in decision support in the field of cultivation.

Lastly, the **application layer,** an intelligence specific that allows the customization of different services based on user requirements and demands. It provides services for users (such as smart homes, smart and precision agriculture, and others).

In agriculture, it provides services of tracking, tracing, identification, and monitoring of an object or things.

19.4 IOT APPLICATION OR ROLE IN AGRICULTURE

Owing to the attributes and importance of IoT services. This section tried by looking at the potential roles agronomists, food processors, entomologists, growers, manufacturers and farmers as well can be derived from the use of IoT technologies, devices, and apps.

Farmers in the field of production, processing, and agricultural sector have started to utilize IoT technologies, as a driving force for sustainable development in a cost-effective manner. IoT embedded sensors or devices are developed in such a way to provides farmers with the needed information for optimizing production processes. Such as precisions in the use of water, fertilizers, seeds, and herbicide on the farm. It as well offers precise and simple ways to improve farming techniques over time. The Internet of Things (IoT) has the potential to transform agriculture in many aspects, be it production, processing, marketing, storing, and distribution. Despite the benefit derived, IoT has some technical limitations ranges from data saving, network, and appropriate hardware to be used, integrity, and data security.

The key roles of the IoT in addressing challenges in agriculture include:

1. **Food Wastage Reduction and Management:** Food waste has been a major challenge affecting human endeavors over the years. The waste usually occurs along the value chain of the food cycle, which includes production, postharvest handling, storage, processing, distribution, and consumption, which is mainly caused by inefficiency in the food supply chains and inadequate information along the phases in the food circle. The use of an IoT food wastage minimization system enables not only real-time information about the status of food products in each phase of the life cycle, but it also predicts its' shelf life for quality and safety assurance. In this way, each chain actor along the value chain can have a prediction of the rest of the shelf life at the moment which is of interest, such that they can decide whether to keep the strategic plan or change it in order to reduce food wastage to it minimal.

2. **Farmer Safety:** IoT enable farmers to get a solution as side effect instead of visiting research institutes or extension worker in persons at the time of emergencies, such as the current COVID-19 pandemic or other plagues. The best functioning of IoT can be described as – when several sensors are attached to the body such as truck, silo, or animal, then it can be used to access the data which are further analyzed using necessary tools. Finally, the farmer can access this for decision-making.

3. **Monitoring of Fruits Quality:** Tomato is a fruit vegetable plant grown mostly in temperate regions of the world. Temperature fluctuation during storage causes loss of moisture content leading to shrinkage and development of microorganisms. However, maintaining proper temperature in the cold chain through real-time monitoring of the storage chamber has been achieved by the use of quality monitoring mote. The IoT mote consists of embedded sensors for measuring humidity, temperature, and the total volatile organic compound.

4. **Farm Security:** Livestock owners in large commercial agriculture can utilize radio frequency identifiers (RFIDs), a wireless bases IoT chip to track their animals be it their locations or their health conditions. The information generated via radio waves helps in identifying sick animals, which further helps in preventing disease spread to the other stocks. IoT embedded devices also help by reducing labor costs through field monitoring for optimal feeding practices.

5. **E-Banking:** In Rural communities, IoT enables technologies that can be utilized in aspects of money depositing, withdrawer, and cash transfer by the farmer(s), who have no access to banks within his or their vicinity. An example is the use of a POS device. Also, in more isolated communities of peoples, rural radio devices can be used to communicate market prices to larger audiences.

6. **Clean Aquaculture and Water Management:** Satellite light radiation devices can be used to detect water pollution in large bodies of water. This functions by utilizing the wavelength of the pollutants to identify what type or form of pollutant it is, which is helpful for good management and clean aquaculture practices.

7. **Disease Detection Management:** IoT devices can be used in monitoring and detection of diseases in crop plants to improve farming efficiency. An example is the use of RFID with disease recognizable ability, which sent information to readers and shared it across the Internet. The farmer or scientist can access this information from a remote place and take necessary actions, automatically crops can be protected from coming diseases.

8. **Automated Irrigation System:** The need for water in agriculture cannot be overemphasized by looking at the growing population. As reported by United Nations World Water Development in 2016, 70% of water consumed worldwide is used in various production practices in agriculture sectors. However, without the use of a smart and efficient approach, agricultural consumption is said to likely to rise by 20% by the year 2050. So, the need for optimal irrigation systems based on information and communication technologies (ICT), sensors, microcontrollers, actuators, etc. arises to minimize water consumption and maximize high yield in crop production.

New irrigation systems come with embedded sensors and actuators to read and control soil moisture parameters. The measured data are transfer via IoT networks to the local servers for data analysis. Many researchers have used several IoT sensors or devices in many developmental projects and programs to analyses different soil profiles. The sensors also provide information about how much water is required by such crops and at what levels to minimize consumption. A smart or automated irrigation system is shown in Figure 19.2. A smart irrigation system is based on embedded sensor technology that enables control of the system remotely without wasting a usable amount of water and ease in the management of crops. Greenhouse or farm examination in order to understand the water requirement level, soil type, humidity, temperature, etc. is the foremost important before any implementation of a smart irrigation system.

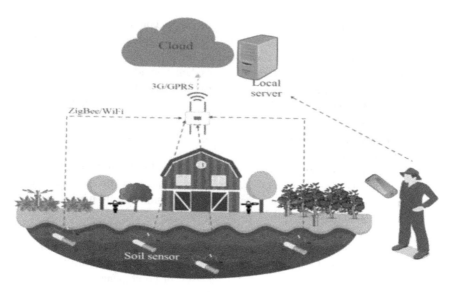

FIGURE 19.2 A modern Irrigation system [1].

9. **Pest Management:** Insect pests of agronomic importance can be early detected and control using IoT sensor-based technologies from attacking crops.

10. **Reduction of Risks:** When farmers' up-to-date information is collected, they can understand what the situation will be in the future, and they can predict some problems that may arise. IoT helps increases agricultural production, reduce manual labour, reduce human error, reduce time and make agricultural activities more efficient.

11. **Weather Forecasting:** Data such as relative humidity, temperature, etc. collected at weather stations over a long period of time can be analyzed to reduce agricultural risk by means of weather forecasting.

12. **Spraying and Fertilizer Application:** Ground-based and aerial-based drones are used in agriculture in order to enhance various agricultural practices: Crop spraying and accurate fertilizer application over large hectares of land to reduce manual labor and for efficiency.

13. **Addressing Food Security Dilemma:** The adoption of IoT-based technologies helps in the enhancement of agricultural production across the continent. Notably, adopting the technologies will not only significantly boost agriculture for food security but also make it more productive, competitive, sustainable, and inclusive. The advent in the field of robotic engineering systems has enabled agri-food ventures to realize better profitability, efficiency, and safety.

14. **Tracking of Harvested Products:** Harvested crops from the farm can be delivered to their destination without harm with the aids of sensor devices. Sensors that use IoT technologies such as GPS and RFID tracking are used to monitor farm products during transportation and storage. This is also

significant for farm products that involve further processing since the buyers can know in advance when the farm products will take place and prepared for the next processing step along the value chain.

15. **Purchasing of Inputs:** Owing to the recent advances in the field of IoT, buying and selling of farm inputs and farm products is made easy through the use of mobile devices. Smartphones equipped with various sensors along with communication network devices such as LTE, Near-Field Communication (NFC), and Wi-Fi, can be used by both consumers and farmers during transactions activities without the use of physical cash, this helps in minimizing the risk of theft as well as fraud prevention.

    In addition, it is a cheaper means for crop producers in rural areas, who have little or no access to financial institutions within their immediate environment to withdraw or deposit cash in the purchasing of seeds, fertilizer, herbicides, feed, implement, and other farming tools.

16. **Wider Marketing opportunity:** The advent of IoT technologies such as smartphones enables wider marketing across different countries. Farmer(s) can easily get links to buyers of his harvest crops without physical barriers and also save costs due to postharvest loss.

17. **Soil Management:** Soil as a vital component of the earth anchored support in the growing and production of crops. Soil parameters such as Moisture content, PH level, porosity, organic matter, etc. can be measured using IoT-enabled devices to guarantee farmers on the best soil for crop production and the soil type that needs proper management before production.

19.5   EXAMPLES OF APPS OF INTERNET OF THINGS (IOT) USED IN AGRICULTURE

- Agrivi mobile app: This app operating system support smartphones of Android, iPad, and iPhone version. It works by providing farmers with easy ways to monitor and analyze farming activities such as tillage, spraying, irrigation, etc. it also helps in data-driven decision making.
- Agrian Mobile app: This application is designed to provide users in the field of crop production with useful up to date information on crop protection and nutritional materials. It helps farmers by giving them insight into the best protection practices as applied to production. The app supported an Android smartphone.
- FarmGraze: In the area of grass and pasture management to save cost and time on farm effectively such mobile app Known as FarmGraze comes in, as it allows you to measure, record, and manage feeds for grazing cattle.
- TractorPal: An invented app with features like an old notebook keep keeping records of inventory for all vehicles including tractors and keeping maintenance records of agricultural machinery and attachments. The app enables users to log small and large machinery in the system and track each item such as purchase date, model year, serial number, purchase price, etc. This app can be operated or used in smartphones such as Android, iPad, iPhone.

- Manure Monitor: This is one of the most widely-used apps in the area of poultry and livestock farming by farmers. It allows users to keep track of important records about their environmental governance.
- Other apps that have found wider application in the field of agriculture include Sirrus use for farm demarcation, MachinaryGuide for visual control of farm inputs, AgFiniti Mobile use for visual mapping, etc.

19.6 CONCLUSION

It is without a doubt that modernization and intensification of the agricultural system could yield the most significant benefits toward supporting food security, employment opportunities, and enhancing inter and intra country trade, innovation, and competitiveness. One of such ways is the adoption and use of IoT embedded technologies referred to as the Internet of Things (IoT). IoT in agriculture refers to the connections of physical components (such as animals, tractors, wire fences, storage, etc.) and the Internet through wireless networks with minimum or no human interventions. It involves the use of devices in turning every entity and action involved in farming into data. The aim of IoT in agriculture is to help farmers by enhancing productivity. It is a well-known fact that the implementation and deployment of these technologies in the agricultural sector brings about greater benefits (ranging from precision agriculture, cost and energy efficiency, and a lot more) to all agricultural stakeholders. The three main components of IoT are the sensing layer, network and data communication layer, and the application layer, which come together to deliver a specific task function. Food waste reduction, farmer safety, farm security, pest management, etc. are some of the key roles played by IoT in agriculture. However, it is expected that in the nearest future, IoT will change every aspect of the agricultural sector.

REFERENCES

1. Tzounis, A., N. Katsoulas, T. Bartzanas and C. Kittas. 2017. Internet of Things in agriculture, recent advances and future challenges, *Biosystems Engineering*, 164, 31–48.
2. Brewster. C., I. Roussaki, N., Kalatzis, K., Doolin and K. Ellis. 2017. IoT in Agriculture: Designing a Europe-Wide Large-Scale Pilot, *IEEE Communications Magazine*.
3. Dlodlo, N., and J. Kalezhi. 2015. The internet of things in agriculture for sustainable rural development, 2015 International Conference on Emerging Trends in Networks and Computer Communications (ETNCC).
4. Abbasi. M., M. H. Yaghmaee and F. Rahnama. 2019. Internet of Things in Agriculture: A Survey, 2019 3rd International Conference on Internet of Things and Applications (IoT).
5. Shafique M.U., W. Ali, and M. Salman. 2019. Rural Development of Pakistan with IoT, *Asian Journal of Research in Computer Science*.

Index